Statistical Thermodynamics and Stochastic Kinetics

An Introduction for Engineers

Presenting the key principles of thermodynamics from a microscopic point of view, this book provides engineers with the knowledge they need to apply thermodynamics and solve engineering challenges at the molecular level. It clearly explains the concepts of entropy and free energy, emphasizing key ideas used in equilibrium applications, whilst stochastic processes, such as stochastic reaction kinetics, are also covered. It provides a classical microscopic interpretation of thermodynamic properties, which is key for engineers, rather than focusing on more esoteric concepts of statistical mechanics and quantum mechanics. Coverage of molecular dynamics and Monte Carlo simulations as natural extensions of the theoretical treatment of statistical thermodynamics is also included, teaching readers how to use computer simulations, and thus enabling them to understand and engineer the microcosm. Featuring many worked examples and over 100 end-of-chapter exercises, it is ideal for use in the classroom as well as for self-study.

YIANNIS N. KAZNESSIS is a Professor in the Department of Chemical Engineering and Materials Science at the University of Minnesota, where he has taught statistical thermodynamics since 2001. He has received several awards and recognitions including the Fulbright Award, the US National Science Foundation CAREER Award, the 3M non-Tenured Faculty Award, the IBM Young Faculty Award, the AIChE Computers and Systems Technology Division Outstanding Young Researcher Award, and the University of Minnesota College of Science and Engineering Charles Bowers Faculty Teaching Award.

This is a well-rounded, innovative textbook suitable for a graduate statistical thermodynamics course, or for self-study. It is clearly written, includes important modern topics (such as molecular simulation and stochastic modeling methods) and has a good number of interesting problems.

Athanassios Z. Panagiotopoulos
Princeton University

Statistical Thermodynamics and Stochastic Kinetics

An Introduction for Engineers

YIANNIS N. KAZNESSIS
University of Minnesota

CAMBRIDGE
UNIVERSITY PRESS

CAMBRIDGE
UNIVERSITY PRESS

University Printing House, Cambridge CB2 8BS, United Kingdom

One Liberty Plaza, 20th Floor, New York, NY 10006, USA

477 Williamstown Road, Port Melbourne, VIC 3207, Australia

314-321, 3rd Floor, Plot 3, Splendor Forum, Jasola District Centre, New Delhi - 110025, India

79 Anson Road, #06-04/06, Singapore 079906

Cambridge University Press is part of the University of Cambridge.

It furthers the University's mission by disseminating knowledge in the pursuit of education, learning and research at the highest international levels of excellence.

www.cambridge.org
Information on this title: www.cambridge.org/9780521765619

First published 2012

A catalogue record for this publication is available from the British Library

Library of Congress Cataloging in Publication data
Kaznessis, Yiannis Nikolaos, 1971 -
Statistical thermodynamics and stochastic kinetics : an introduction
for engineers / Yiannis Nikolaos Kaznessis.
p.   cm.
Includes index.
ISBN 978-0-521-76561-9
1. Statistical thermodynamics.   2. Stochastic processes.
3. Molucular dynamics–Simulation methods.   I. Title.
TP155.2.T45K39   2012
536´.7–dc23   2011031548

ISBN 978-0-521-76561-9 Hardback

To my beloved wife, Elaine

Contents

Appendices

Acknowledgments

I am grateful for the contributions that many people have made to this book. Ed Maggin was the first to teach me Statistical Thermodynamics and his class notes were always a point of reference. The late Ted H. Davis gave me encouragement and invaluable feedback. Dan Bolintineanu and Thomas Jikku read the final draft and helped me make many corrections. Many thanks go to the students who attended my course in Statistical Thermodynamics and who provided me with many valuable comments regarding the structure of the book. I also wish to thank the students in my group at Minnesota for their assistance with making programs available on *sourceforge.net*. In particular, special thanks go to Tony Hill who oversaw the development and launch of the stochastic reaction kinetics algorithms. Finally, I am particularly thankful for the support of my wife, Elaine.

1

Introduction

1.1 Prologue

Engineers learn early on in their careers how to harness energy from nature, how to generate useful forms of energy, and how to transform between different energy forms. Engineers usually first learn how to do this in thermodynamics courses.

There are two fundamental concepts in thermodynamics, energy, E, and entropy, S. These are taught axiomatically in engineering courses, with the help of the two laws of thermodynamics:
(1) energy is always conserved, and
(2) the entropy difference for any change is non-negative.

Typically, the first law of thermodynamics for the energy of a system is cast into a balance equation of the form:

$$\left\{\begin{array}{c} \text{change of energy in the system} \\ \text{between times } t_1 \text{ and } t_2 \end{array}\right\} = \left\{\begin{array}{c} \text{energy that entered the system} \\ \text{between times } t_1 \text{ and } t_2 \end{array}\right\} -$$

$$\left\{\begin{array}{c} \text{energy that exited the system} \\ \text{between times } t_1 \text{ and } t_2 \end{array}\right\} + \left\{\begin{array}{c} \text{energy generated in the system} \\ \text{between times } t_1 \text{ and } t_2 \end{array}\right\}.$$

$$(1.1)$$

The second law of thermodynamics for the entropy of a system can be presented through a similar balance, with the generation term never taking any negative values. Alternatively, the second law is presented with an inequality for the entropy, $\Delta S \geq 0$, where ΔS is the change of entropy of the system for a well-defined change of the system's state.

These laws have always served engineering disciplines well. They are adequate for purposes of engineering distillation columns, aircraft engines, power plants, fermentation reactors, or other large, macroscopic systems and processes. Sound engineering practice is inseparable from understanding the first principles underlying physical phenomena and processes, and the two laws of thermodynamics form a solid core of this understanding.

Macroscopic phenomena and processes remain at the heart of engineering education, yet the astonishing recent progress in fields like nanotechnology and genetics has shifted the focus of engineers to the microcosm. Thermodynamics is certainly applicable at the microcosm, but absent from the traditional engineering definitions is a molecular interpretation of energy and entropy. Understanding thermodynamic behavior at small scales can then be elusive.

The goal of this book is to present thermodynamics from a microscopic point of view, introducing engineers to the body of knowledge needed to apply thermodynamics and solve engineering challenges at the molecular level. Admittedly, this knowledge has been created in the physical and chemical sciences for more than one hundred years, with statistical thermodynamics. There have been hundreds of books published on this subject, since Josiah Willard Gibbs first developed his ensemble theory in the 1880s and published the results in a book in 1902. What then could another textbook have to offer?

I am hoping primarily three benefits:

1. A microscopic interpretation of thermodynamic concepts that engineers will find appropriate, one that does not dwell in the more esoteric concepts of statistical thermodynamics and quantum mechanics. I should note that this book does not shy away from mathematical derivations and proofs. I actually believe that sound mathematics is inseparable from physical intuition. But in this book, the presentation of mathematics is subservient to physical intuition and applicability and not an end in itself.

2. A presentation of molecular dynamics and Monte Carlo simulations as natural extensions of the theoretical treatment of statistical thermodynamics. I philosophically subscribe to the notion that computer simulations significantly augment our natural capacity to study and understand the natural world and that they are as useful and accurate as their underlying theory. Solidly founded on the theoretical concepts of statistical thermodynamics, computer simulations can become a potent instrument for assisting efforts to understand and engineer the microcosm.

3. A brief coverage of stochastic processes in general, and of stochastic reaction kinetics in particular. Many dynamical systems of scientific and technological significance are not at the thermodynamic limit (systems with very large numbers of particles). Stochasticity then emerges as an important feature of their dynamic behavior. Traditional continuous-deterministic models, such as reaction rate

ordinary differential equations for reaction kinetics, do not capture the probabilistic nature of small systems. I present the theory for stochastic processes and discuss algorithmic solutions to capture the probabilistic nature of systems away from the thermodynamic limit.

To provide an outline of the topics discussed in the book, I present a summary of the salient concepts of statistical thermodynamics in the following section.

1.2 If we had only a single lecture in statistical thermodynamics

The overarching goal of classical statistical thermodynamics is to explain thermodynamic properties of matter in terms of atoms. Briefly, this is how:

Consider a system with N identical particles contained in volume V with a total energy E. Assume that N, V, and E are kept constant. We call this an NVE system (Fig. 1.1). These parameters uniquely define the macroscopic state of the system, that is all the rest of the thermodynamic properties of the system are defined as functions of N, V, and E. For example, we can write the entropy of the system as a function $S = S(N, V, E)$, or the pressure of the system as a function $P = P(N, V, E)$. Indeed, if we know the values of N, V, and E for a single-component, single-phase system, we can in principle find the values of the enthalpy H, the Gibbs free energy G, the Helmholtz free energy A, the chemical potential μ, the entropy S, the pressure P, and the temperature T. In Appendix B, we summarize important elements of thermodynamics, including the fundamental relations between these properties.

Figure 1.1 System with N particles contained in volume V with a total energy E.

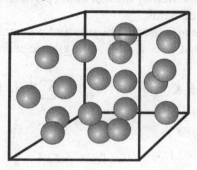

A fundamentally important concept of statistical thermodynamics is the microstate of a system. We define a microstate of a system by the values of the positions and velocities of all the N particles. We can concisely describe a microstate with a $6N$-dimensional vector

$$\underline{X} = (\underline{r}_1, \underline{r}_2, \ldots, \underline{r}_N, \underline{\dot{r}}_1, \underline{\dot{r}}_2, \ldots, \underline{\dot{r}}_N). \tag{1.2}$$

In Eq. 1.2, $\underline{r}_i$ are the three position coordinates and $\underline{\dot{r}}_i$ are the three velocity coordinates of particle i, respectively, with $i = 1, 2, \ldots, N$. By definition, $\underline{\dot{r}}_i = d\underline{r}_i/dt$. Note that the positions and the velocities of atoms do not depend on one another.

An important postulate of statistical thermodynamics is that each macroscopic property M of the system (for example the enthalpy H, or the pressure P) at any time t is a function of the positions and velocities of the N particles at t, i.e., $M(t) = M(\underline{X}(t))$. Then, any observed, experimentally measured property M_{observed} is simply the time average of instantaneous values $M(t)$,

$$M_{\text{observed}} = \langle M \rangle = \lim_{T \to \infty} \frac{1}{T} \int_0^T M(\underline{X}(t))dt, \tag{1.3}$$

where T is the time of the experimental measurement.

Equation (1.3) provides a bridge between the observable macroscopic states and the microscopic states of any system. If there were a way to know the microscopic state of the system at different times then all thermodynamic properties could be determined. Assuming a classical system of point-mass particles, Newtonian mechanics provides such a way. We can write Newton's second law for each particle i as follows:

$$m_i \underline{\ddot{r}}_i = \underline{F}_i, \tag{1.4}$$

where m_i is the mass of particle i, $\underline{\ddot{r}}_i = d^2\underline{r}_i/dt^2$, and $\underline{F}_i$ is the force vector on particle i, exerted by the rest of the particles, the system walls, and any external force fields.

We can define the microscopic kinetic and potential energies, K and U, respectively so that $E = K + U$. The kinetic energy is

$$K = K(\underline{\dot{r}}_1, \underline{\dot{r}}_2, \ldots, \underline{\dot{r}}_N) = \sum_{i=1}^{N} \frac{1}{2} m_i \dot{r}_i^2. \tag{1.5}$$

The potential energy is

$$U = U(\underline{r}_1, \underline{r}_2, \ldots, \underline{r}_N), \tag{1.6}$$

so that (for conservative systems)

$$\underline{F}_i = -\frac{\partial U}{\partial \underline{r}_i}.$$
(1.7)

Albert Einstein attempted to infer the laws of thermodynamics from Newtonian mechanics for systems with large but finite degrees of freedom. In principle, a set of initial conditions at $t = 0$, $\underline{X}(0)$, would suffice to solve the second law of motion for each particle, determine $\underline{X}(t)$ and through Eq. (1.3) determine thermodynamic properties. Einstein was, however, unsuccessful in his quest. A simple reason is that it is not practically feasible to precisely determine the initial microscopic state of a system with a large number of particles N, because it is not possible to conduct $6N$ independent experiments simultaneously.

The impossibility of this task notwithstanding, even if the initial conditions of a system could be precisely determined in a careful experiment at $t = 0$, the solution of $6N$ equations of motion in time is not possible for large numbers of particles. Had Einstein had access to the super-computing resources available to researchers today, he would still not be able to integrate numerically the equations of motion for any system size near $N = 10^{23}$. To appreciate the impossibility of this task, assume that a computer exists that can integrate for one time step 10 000 coupled ordinary differential equations in one wall-clock second. This computer would require 10^{20} seconds to integrate around 10^{24} equations for this single time step. With the age of the universe being, according to NASA, around 13.7 billion years, or around 432×10^{15} seconds, the difficulty of directly connecting Newtonian mechanics to thermodynamics becomes apparent.

Thankfully, Josiah Willard Gibbs* developed an ingenious conceptual framework that connects the microscopic states of a system to macroscopic observables. He accomplished this with the help of the concept of phase space (Fig. 1.2). For a system with N particles, the phase space is a $6N$ dimensional space where each of the $6N$ orthogonal axes corresponds to one of the $6N$ degrees of freedom, i.e., the positions and velocities of the particles. Each point in phase space is identified by a vector

$$\underline{X} = (\underline{r}_1, \underline{r}_2, \ldots, \underline{r}_N, \underline{\dot{r}}_1, \underline{\dot{r}}_2, \ldots, \underline{\dot{r}}_N),$$
(1.8)

---

* It is noteworthy that Gibbs earned a Ph.D. in Engineering from Yale in 1863. Actually, his was the first engineering doctorate degree awarded at Yale. Gibbs had studied Mathematics and Latin as an undergraduate and stayed at Yale for all of his career as a Professor in Mathematical Physics.

or equivalently by a vector

$$\underline{X} = (\underline{r}_1, \underline{r}_2, \ldots, \underline{r}_N, \underline{p}_1, \underline{p}_2, \ldots, \underline{p}_N), \tag{1.9}$$

where $\underline{p}_i = m_i \underline{\dot{r}}_i$, is the momentum of particle i.

Consequently, each point in phase space represents a microscopic state of the system. For an NVE system the phase space is finite, since no position axis can extend beyond the confines of volume V and no momentum axis can extend beyond a value that yields the value of the total kinetic energy.

In classical mechanics the phase space is finite, of size Σ, but because it is continuous, the number of microscopic states is infinite. For each state identified with a point $\underline{X}$, a different state can be defined at $\underline{X} + d\underline{X}$, where $d\underline{X}$ is an infinitesimally small distance in $6N$ dimensions.

Thanks to quantum mechanics, we now know that this picture of a continuous phase space is physically unattainable. Werner Heisenberg's uncertainty principle states that the position and momentum of a particle cannot be simultaneously determined with infinite precision. For a particle confined in one dimension, the uncertainties in the position, Δx, and momentum, Δp, cannot vary independently: $\Delta x \Delta p \geq h/4\pi$, where $h = 6.626 \times 10^{-34}$ m^2 kg/s is Planck's constant.

The implication for statistical mechanics is significant. What the quantum mechanical uncertainty principle does is simply to discretize the phase space (Fig. 1.3). For any NVE system, instead of an infinite number of possible microscopic states, there is a finite number of microscopic states corresponding to the macroscopic NVE system. Let us call this number Ω and write $\Omega(N, V, E)$ to denote that it is determined by the macroscopic state.

Figure 1.2 Phase space Γ. Each microscopic state of a macroscopic NVE system is represented by a single point in $6N$ dimensions.

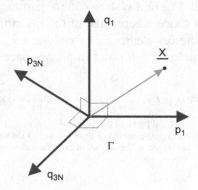

Another fundamental postulate of statistical thermodynamics is that all these Ω microscopic states have the same probability of occurring. This probability is then

$$P = 1/\Omega. \tag{1.10}$$

Ludwig Boltzmann showed around the same time as Gibbs that the entropy of an NVE system is directly related to the number of microscopic states Ω. Gibbs and Boltzmann were thus able to provide a direct link between microscopic and macroscopic thermodynamics, one that proved to be also useful and applicable. The relation between entropy $S(N, V, E)$ and the number of microscopic states $\Omega(N, V, E)$ has been determined by numerous different methods. We will present a concise one that Einstein proposed:

1. Assume there generally exists a specific function that relates the entropy of an NVE system to the number of microscopic states that correspond to this NVE macroscopic state. The relation can be written as

$$S = \phi(\Omega). \tag{1.11}$$

2. Consider two independent systems A and B. Then

$$S_A = \phi(\Omega_A), \tag{1.12}$$

and

$$S_B = \phi(\Omega_B). \tag{1.13}$$

3. Consider the composite system of A and B. Call it system AB. Since entropy is an extensive property, the entropy of the composite system is

$$S_{AB} = \phi(\Omega_{AB}) = S_A + S_B = \phi(\Omega_A) + \phi(\Omega_B). \tag{1.14}$$

Figure 1.3 The available phase space to any macroscopic state is an ensemble of discrete microscopic states. The size of the available phase space is Σ, and the number of microscopic states is Ω.

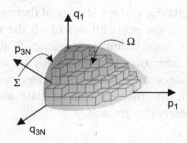

4. Since the systems are independent, the probability of the composite system being in a particular microscopic state is equal to the product of probabilities that systems A and B are in their respective particular microscopic state, i.e.,

$$P_{AB} = P_A P_B. \tag{1.15}$$

Therefore the number of microscopic states of the composite system can be written as

$$\Omega_{AB} = \Omega_A \Omega_B. \tag{1.16}$$

5. Combining the results in the two previous steps,

$$\phi(\Omega_{AB}) = \phi(\Omega_A \Omega_B) = \phi(\Omega_A) + \phi(\Omega_B). \tag{1.17}$$

The solution of this equation is

$$\phi(\Omega) = k_B \ln(\Omega), \tag{1.18}$$

and thus

$$S = k_B \ln(\Omega), \tag{1.19}$$

where $k_B = 1.38065 \times 10^{-23}\ \mathrm{m^2 kg\, s^{-2} K^{-1}}$ is Boltzmann's constant.

This equation, which is called Boltzmann's equation, provides a direct connection between microscopic and macroscopic properties of matter. Importantly, the entropy of NVE systems is defined in a way that provides a clear physical interpretation.

Looking at the phase space not as a succession in time of microscopic states that follow Newtonian mechanics, but as an ensemble of microscopic states with probabilities that depend on the macroscopic state, Gibbs and Boltzmann set the foundation of statistical thermodynamics, which provides a direct connection between classical thermodynamics and microscopic properties.

This has been accomplished not only for NVE systems, but for NVT, NPT, and μVT systems among others. Indeed, for any system in an equilibrium macroscopic state, statistical thermodynamics focuses on the determination of the probabilities of all the microscopic states that correspond to the equilibrium macrostate. It also focuses on the enumeration of these microscopic states. With the information of how many microscopic states correspond to a macroscopic one and of what their probabilities are, the thermodynamic state and behavior of the system can be completely determined.

Remembering from thermodynamics that

$$dE = TdS - PdV + \mu dN, \tag{1.20}$$

we can write, for the NVE system

$$\left.\frac{\partial S}{\partial E}\right|_{N,V} = \frac{1}{T}, \tag{1.21}$$

or

$$\left.\frac{\partial \ln(\Omega)}{\partial E}\right|_{N,V} = \frac{1}{k_B T}. \tag{1.22}$$

Similarly,

$$\left.\frac{\partial \ln(\Omega)}{\partial V}\right|_{N,E} = \frac{P}{k_B T}, \tag{1.23}$$

and

$$\left.\frac{\partial \ln(\Omega)}{\partial N}\right|_{E,V} = -\frac{\mu}{k_B T}. \tag{1.24}$$

In this book I present the theory for enumerating the microscopic states of equilibrium systems and determining their probabilities. I then discuss how to use this knowledge to derive thermodynamic properties, using Eqs. 1.21–1.24, or other similar ones for different ensembles.

As an example, consider an ideal gas of N particles, in volume V, with energy E. The position of any of these non-interacting particles is independent of the positions of the rest of the particles. We discuss in Chapter 4 that in this case we can enumerate the microscopic states. In fact we find that

$$\Omega(N, V, E) \propto V^N. \tag{1.25}$$

Using Eq. 1.23 we can then write

$$\frac{P}{k_B T} = \frac{N}{V}, \tag{1.26}$$

and rearranging

$$PV = Nk_B T. \tag{1.27}$$

We can show that the Boltzmann constant is equal to the ratio of the ideal gas constant over the Avogadro number, $k_B = R/N_A$. Then for ideal gases

$$PV = nRT, \tag{1.28}$$

where n is the number of moles of particles in the system.

First stated by Benoît Paul Emile Clapeyron in 1834, the ideal gas law, an extraordinary and remarkably simple equation that has since guided understanding of gas thermodynamics, was originally derived empirically. With statistical thermodynamics the ideal gas law is derived theoretically from simple first principles and statistical arguments.

I discuss how other equations of state can be derived theoretically using information about the interactions at the atomic level. I do this analytically for non-ideal gases, liquids, and solids of single components of monoatomic and of diatomic molecules. I then introduce computer simulation techniques that enable us numerically to connect the microcosm with the macrocosm for more complex systems, for which analytical solutions are intractable.

In Chapter 2, I present the necessary elements of probability and combinatorial theory to enumerate microscopic states and determine their probability. I assume no prior exposure to statistics, which is regretfully true for most engineers.

I then discuss, in Chapter 3, the classical mechanical concepts required to define microscopic states. I introduce quantum mechanics in order to discuss the notion of a discrete phase space. In Chapter 4, I introduce the classical ensemble theory, placing emphasis on the NVE ensemble.

In Chapter 5, I define the canonical NVT ensemble. In Chapter 6, fluctuations and the equivalence of various ensembles is presented. Along the way, we derive the thermodynamic properties of monoatomic ideal gases.

Diatomic gases, non-ideal gases, liquids, crystals, mixtures, reacting systems, and polymers are discussed in Chapters 7–11.

I present an introduction to non-equilibrium thermodynamics in Chapter 12, and stochastic processes in Chapter 13.

Finally, in Chapters 14–18, I introduce elements of Monte Carlo, molecular dynamics and stochastic kinetic simulations, presenting them as the natural, numerical extension of statistical mechanical theories.

Elements of probability and combinatorial theory

αριθμω δε τα παντα επεοικεν

Pythagoras (570–495 BC)

2.1 Probability theory

There are experiments with more than one outcome for any trial. If we do not know which outcome will result in a given trial, we define outcomes as random and we assign a number to each outcome, called the probability. We present two distinct definitions of probability:

1. **Classical probability.** Given W possible simple outcomes to an experiment or measurement, the classical probability of a simple event E_i is defined as

$$P(E_i) = 1/W. \tag{2.1}$$

Example 2.1

If the experiment is tossing a coin, there are $W = 2$ possible outcomes: $E_1 = $ "*heads*," $E_2 = $ "*tails*." The probability of each outcome is

$$P(E_i) = 1/2. \quad i = 1, 2 \tag{2.2}$$

2. **Statistical probability.** If an experiment is conducted N times and an event E_i occurs n_i times ($n_i \le N$), the statistical probability of this event is

$$P(E_i) = \lim_{N \to \infty} \frac{n_i}{N}. \tag{2.3}$$

The statistical probability converges to the classical probability when the number of trials is infinite. If the number of trials is small, then the value of the statistical probability fluctuates. We show later in this chapter that the magnitude of fluctuations in the value of $P(E_i)$ is inversely proportional to $\sqrt{N}$.

2.1.1 Useful definitions

1. The value of all probabilities is always bounded: $0 \le P(E_i) \le 1, \forall E_i$.
2. The probability that two events E_1 and E_2 occur is called the joint probability and is denoted by $P(E_1, E_2)$.
3. Two events E_1 and E_2, with probabilities $P(E_1)$ and $P(E_2)$, respectively, are called *mutually exclusive*, when the probability that one of these events occurs is

$$P = P(E_1) + P(E_2). \qquad (2.4)$$

Example 2.2
Consider a deck of cards. The probability of drawing the ace of spades is $P(\text{“ace of spades”}) = 1/52$. The probability of drawing a ten is $P(\text{“ten”}) = 4/52$. The two events, drawing the ace of spades or a ten, are mutually exclusive, because they cannot both occur in a single draw. The probability of either one occurring is equal to the sum of event probabilities, $P = P(\text{“ace of spades”}) + P(\text{“ten”}) = 5/52$.

4. Two events E_1 and E_2, with respective probabilities $P(E_1)$ and $P(E_2)$, are called *independent* if the joint probability of both of them occurring is equal to the product of the individual event probabilities

$$P(E_1, E_2) = P(E_1)P(E_2). \qquad (2.5)$$

Example 2.3
Consider two separate decks of cards. Drawing a ten from both has a probability $P(\text{“ten”}, \text{“ten”}) = P(\text{“ten”})P(\text{“ten”}) = (4/52)^2 = 0.0059$.

5. The *conditional* probability that an event E_2 occurs provided that event E_1 has occurred is denoted by $P(E_2|E_1)$. It is related to the joint probability through

$$P(E_1, E_2) = P(E_1)P(E_2|E_1). \qquad (2.6)$$

Example 2.4
Consider a single deck of cards. The probability that a ten is drawn first and the ace of spades is drawn second in two consecutive draws is $P = P(\text{“ten”})P(\text{“ace of spades”}|\text{“ten”}) = (4/52)(1/51)$.

6. For two independent events

$$P(E_2|E_1) = P(E_2). \qquad (2.7)$$

7. If $(E_1 + E_2)$ denotes either E_1 or E_2 or both, then

$$P(E_1 + E_2) = P(E_1) + P(E_2) - P(E_1, E_2). \tag{2.8}$$

For mutually exclusive events, $P(E_1, E_2) = 0$. Then,

$$P(E_1 + E_2) = P(E_1) + P(E_2). \tag{2.9}$$

2.1.2 Probability distributions

We present two types of probability distribution: discrete and continuous.

Discrete distributions

Consider a variable X, which can assume discrete values $X_1, X_2, \ldots, X_K$ with respective probabilities $P(X_1), P(X_2), \ldots, P(X_K)$.

Note that X is called a random variable. The function $P(X)$ is a discrete probability distribution function. By definition

$$\sum_{k=1}^{K} P(X_k) = 1, \tag{2.10}$$

where the sum runs over all K possible discrete values of X.

Example 2.5

Consider a single die. There are six outcomes with the same probability:

X	1	2	3	4	5	6
P(X)	1/6	1/6	1/6	1/6	1/6	1/6

This is a simple example of a uniform distribution (Fig. 2.1).

Figure 2.1 Uniform discrete probability distribution. X is the face value of a fair die.

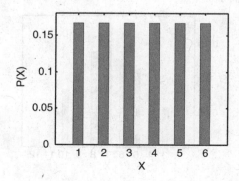

Example 2.6

Consider two dice. If X is equal to the sum of the face values that the two dice assume in a roll, we can write (Fig. 2.2):

X sum of points	2	3	4	5	6	7	8	9	10	11	12
P(X)	1/36	2/36	3/36	4/36	5/36	6/36	5/36	4/36	3/36	2/36	1/36

Continuous distributions

Consider a random variable x, which can assume a continuous set of values in the interval $[a, b]$ with frequencies $f(x)$.

We define the probability density function of x as the relative frequency

$$\rho(x)\,dx = \frac{f(x)\,dx}{\int_a^b f(x)\,dx},\qquad(2.11)$$

so that

$$\int_a^b \rho(x)\,dx = 1.\qquad(2.12)$$

The limits $[a, b]$ can extend to infinity.

Example 2.7

Consider a random variable x equal to the weight of a newborn baby. Since practically the weight can change in a continuous fashion, we can measure the frequencies $f(x)$ in a large sample of newborn babies and define the probability density with Eq. 2.11.

Figure 2.2 Non-uniform discrete probability distribution. X is the sum of the face values of two fair dice.

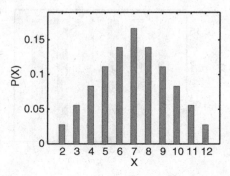

2.1.3 Mathematical expectation

1. The mathematical expectation of a discrete random variable X is defined as

$$E(X) = \langle X \rangle = \frac{P(X_1)X_1 + P(X_2)X_2 + \cdots + P(X_K)X_K}{P(X_1) + P(X_2) + \cdots + P(X_K)}, \quad (2.13)$$

or more concisely

$$E(X) = \langle X \rangle = \sum_{k=1}^{K} X_k P(X_k). \quad (2.14)$$

Example 2.8
Consider X being the face value of a rolling die. The expectation is $\langle X \rangle = 3.5$.

2. The mathematical expectation of a continuous random variable x is defined as

$$E(x) = \frac{\int x \rho(x)\,dx}{\int \rho(x)\,dx}, \quad (2.15)$$

or

$$E(x) = \int x \rho(x)\,dx. \quad (2.16)$$

Note that for simplicity, we will not write the limits of integration from now on.

3. Generally for any function $g(X)$ of a discrete random variable, or $g(x)$ of a continuous random variable, the expectation of the function can be calculated as

$$E(g(X)) = \langle g(X) \rangle = \sum_{k=1}^{K} P(X_k) g(X_k) \quad (2.17)$$

for the case of a discrete variable, and

$$E(g(x)) = \langle g(x) \rangle = \int g(x) \rho(x)\,dx \quad (2.18)$$

for the case of a continuous random variable.

2.1.4 Moments of probability distributions

Probability distributions can be described by their moments. Consider the probability distribution of a random variable x. The mth moment of

the distribution is defined as

$$\mu_m = \langle (x - \langle x \rangle)^m \rangle. \tag{2.19}$$

1. The zeroth moment is $\mu_0 = 1$.
2. The first moment is $\mu_1 = 0$.
3. The second moment μ_2 is also called the variance and denoted by V:

$$V = \mu_2 = \langle (x - \langle x \rangle) \rangle^2. \tag{2.20}$$

The standard deviation is defined by

$$\sigma = \sqrt{\mu_2}. \tag{2.21}$$

4. The third moment is related to the skewness of a distribution, a_3, a measure of asymmetry, as follows

$$a_3 = \mu_3 / \sigma^3. \tag{2.22}$$

5. The fourth moment is related to the kurtosis of a distribution, a_4, a measure of narrowness, as follows

$$a_4 = \mu_4 / \sigma^4. \tag{2.23}$$

2.1.5 Gaussian probability distribution

The normal or Gaussian probability distribution is one we will encounter often (Fig. 2.3). It is defined as follows:

$$\rho(x) = \frac{1}{(2\pi\sigma)^{1/2}} \exp\left[-\frac{(x - \langle x \rangle)^2}{2\sigma^2} \right]. \tag{2.24}$$

Figure 2.3 Gaussian probability distributions with different mean, μ, and standard deviation, σ, values.

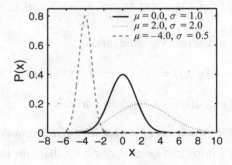

This is an important distribution that is frequently observed in natural phenomena. We will see later in this book how the Gaussian distribution emerges naturally for many physical phenomena.

Johann Carl Friedrich Gauss, considered by many the greatest mathematician ever to exist, introduced this distribution in the early nineteenth century, although the normal distribution was first described by Abraham de Moivre in 1733.*

A well-known example of a Gaussian distribution is the celebrated Maxwell–Boltzmann distribution of velocities of gas particles. In one dimension, we write

$$f(u_x) = \left(\frac{m}{2k_B T}\right)^{3/2} \exp\left(-\frac{m u_x^2}{2k_B T}\right). \tag{2.25}$$

We derive this equation and discuss it in detail in Chapter 5.

2.2 Elements of combinatorial analysis

A fundamental principle of combinatorial analysis is that if an event E_1 can occur in n_1 ways and an event E_2 can occur in n_2 different ways, then the number of ways both can occur is equal to the product $n_1 n_2$.

2.2.1 Arrangements

The number of ways to arrange n dissimilar objects is

$$n! = 1 \times 2 \times \cdots \times (n-1) \times n. \tag{2.26}$$

Example 2.9
Consider three dissimilar objects: A, B, and C. The number of arrangements is equal to $1 \times 2 \times 3 = 6$:

$$(ABC, \quad ACB, \quad BAC, \quad BCA, \quad CAB, \quad CBA).$$

If there are n objects of which p are identical, then the number of arrangements is $n!/p!$. Generally if n objects consist of p objects of one kind, q objects of another kind, r of a third kind, etc. then the number of arrangements is $n!/(p!q!r!\dots)$.

---

* This is an instance of Stephen Stigler's law, which states that "No scientific discovery is named after its original discoverer." According to Wikipedia, Joel Cohen surmised that "Stiglers law was discovered many times before Stigler named it."

2.2.2 Permutations

Permutation is the number of ways we can choose and arrange r objects out of a total of n objects (Fig. 2.4). Denoted by $_nP_r$, permutations are equal to

$$_nP_r = \frac{n!}{(n-r)!}. \tag{2.27}$$

Example 2.10
Consider the four letters A, B, C, D. The number of ways of choosing and arranging two is

$$_4P_2 = \frac{4!}{(4-2)!} = \frac{24}{2} = 12. \tag{2.28}$$

These 12 arrangements are

$$(AB, \; AC, \; AD, \; BA, \; BC, \; BD, \; CA, \; CB, \; CD, \; DA, \; DB, \; DC).$$

2.2.3 Combinations

If the order of object arrangement is not important, then the number of ways of choosing and arranging r objects out of a total of n is called combinations, denoted with $_nC_r$ and calculated as follows:

$$_nC_r = \frac{n!}{r!\,(n-r)!}. \tag{2.29}$$

Example 2.11
In the previous example, outcome AB is equivalent to, or indistinguishable from, outcome BA. The number of combinations of two letters out

Figure 2.4 There are six ways to place three distinguishable balls in three boxes when only one ball is allowed per box (A). There is only one way to place three indistinguishable balls (B).

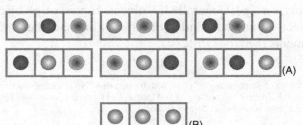

(A)

(B)

of four is then

$$_4C_2 = \frac{4!}{2!2!} = 6. \qquad (2.30)$$

2.3 Distinguishable and indistinguishable particles

Consider a system of N particles in volume V. If the macroscopic state of the system does not change when particles exchange their positions and velocities, then particles are called indistinguishable. Otherwise, they are called distinguishable.

Example 2.12

Consider three balls that are distinguishable. For example, one is red, another is blue, and the third is green. In how many ways can these balls be placed in three boxes, A, B, and C, if only one ball is allowed per box? The answer is obviously $3! = 6$ different ways. If, on the other hand, the balls are indistinguishable (e.g., they are all blue) then there is only one way to place these balls one in each box (Fig. 2.4).

Similarly, we can find that there are $4! = 24$ ways to place four distinguishable balls in four different boxes, with at most one ball per box, or that there are $5! = 120$ ways to place five distinguishable balls in five different boxes. In general, there are $M!$ ways to place $M!$ distinguishable balls in M boxes, with at most one ball in each box.

If the balls are indistinguishable we can calculate the number of different ways to place three balls in four boxes to be just four, and the number of ways to place three balls in five boxes to be just 20.

In general, there are $M!/N!$ ways to place N indistinguishable balls in M boxes, with at most one ball in each box ($M \geq N$ must still be true). We will see that the term $1/N!$ appears in calculations of microscopic states, changing the value of thermodynamic properties.

Example 2.13

Consider a different case, where more than one particle can be placed in a box. For example, consider ten distinguishable particles and three boxes. How many ways are there to place three particles in one box, five particles in another box, and two particles in a third box?

The first three particles can be chosen with the following number of ways:

$$_{10}C_3 = \frac{10!}{3!(10-3)!}. \tag{2.31}$$

The second group of five can then be chosen with the following number of ways:

$$_7C_5 = \frac{7!}{5!\,2!}. \tag{2.32}$$

The final group of two particles can be chosen with the following number of ways:

$$_2C_2 = \frac{2!}{2!\,0!} = 1. \tag{2.33}$$

The total number of ways is

$$W = {}_{10}C_3\,{}_7C_5\,{}_2C_2 = \frac{10!}{3!\,5!\,2!}. \tag{2.34}$$

In general, the number of ways we can arrange N distinguishable items in a number of different states, with n_i in a state i is equal to

$$W = \frac{N!}{n_1!n_2!\ldots}. \tag{2.35}$$

2.4 Stirling's approximation

The value of $N!$ is practically impossible to calculate for large N. For example, if N is the number of molecules of a macroscopic system containing 1 mole, there is no direct way to calculate the term $6.023 \times 10^{23}!$.

James Stirling (1692–1770) devised the following approximation to compute $\ln N!$:

$$N! \approx \sqrt{2\pi N}N^N e^{-N} \tag{2.36}$$

or

$$\ln N! \approx \left(N + \frac{1}{2}\right)\ln N - N + \frac{1}{2}\ln(2\pi), \tag{2.37}$$

and finally,

$$\ln N! \approx N \ln N - N. \tag{2.38}$$

A simple way to see how this approximation emerges is the following:

$$\ln N! = \sum_{m=1}^{N} \ln m \approx \int_{1}^{N} \ln x\, dx = N \ln N - N. \qquad (2.39)$$

Equations 2.36 and 2.38 are called Stirling's formula. We will use them both interchangeably.

2.5 Binomial distribution

Consider p as the probability an event will occur in a single trial (success) and q as the probability this same event will not occur in a single trial (failure). Of course, $p + q = 1$.

The probability that the event will occur X times in N trials is given by the binomial distribution (Fig. 2.5)

$$P(X) = \binom{N}{X} p^X q^{N-X} = \frac{N!}{(N-X)!\,X!} p^X q^{N-X}. \qquad (2.40)$$

Example 2.14
The probability of getting two heads in six coin tosses is

$$\binom{6}{2} \left(\frac{1}{2}\right)^2 \left(\frac{1}{2}\right)^{6-2} = \frac{6!}{2!\,4!} \left(\frac{1}{2}\right)^6 = \frac{15}{64}. \qquad (2.41)$$

Figure 2.5 Binomial probability distributions with different values of success probabilities, p, and number of trials, N.

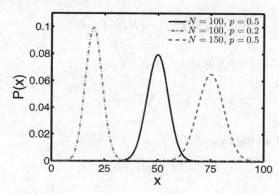

According to the binomial expansion

$$\sum_{X=0}^{N} P(X) = [p+q]^N = 1. \tag{2.42}$$

The mean and the variance of a random variable X that is binomially distributed can be determined as follows. First, consider a new, auxiliary variable t^X.

The expectation value of t^X is

$$E(t^X) = \sum_{X=0}^{N} P(X)t^X = \sum_{X=0}^{N} t^X \binom{N}{X} p^X q^{N-X} \tag{2.43}$$

or

$$E(t^X) = \sum_{X=0}^{N} t^X \frac{N!}{(N-X)!\, X!} p^X q^{N-X}, \tag{2.44}$$

and

$$E(t^X) = q^N + Npq^{N-1}t + \cdots + p^N t^N. \tag{2.45}$$

Finally,

$$E(t^X) = (q + pt)^N \tag{2.46}$$

or

$$\sum_{X=0}^{N} P(X)t^X = (q + pt)^N. \tag{2.47}$$

Differentiating with respect to t yields

$$\sum_{X=0}^{N} P(X)Xt^{X-1} = N(q + pt)^{N-1}p. \tag{2.48}$$

The expectation of X is defined as

$$E(X) = \sum_{X=0}^{N} XP(X). \tag{2.49}$$

Equation 2.48 then yields for $t = 1$

$$E(X) = Np. \tag{2.50}$$

Similarly, the variance of the binomial distribution can be determined as

$$\sigma^2 = E[(X - E(X))^2] = Npq \tag{2.51}$$

using

$$E(X^2) = N(N - 1)p^2. \tag{2.52}$$

2.6 Multinomial distribution

If events $E_1, E_2, \ldots, E_K$ can occur with probabilities $P_1, P_2, \ldots, P_K$, respectively, the probability that events $E_1, E_2, \ldots, E_K$ will occur $X_1, X_2, \ldots, X_K$ times respectively is given by the multinomial distribution

$$P(X_1, X_2, \ldots, X_K) = \frac{N!}{X_1! X_2! \ldots X_K!} P_1^{X_1} P_2^{X_2} \ldots P_K^{X_K}. \tag{2.53}$$

2.7 Exponential and Poisson distributions

The exponential distribution function

$$P(x) = \lambda \exp(-\lambda X), \tag{2.54}$$

describes time events that occur at a constant average rate, λ.

The Poisson distribution

$$P(X) = \frac{\lambda^X}{X!} e^{-\lambda}, \tag{2.55}$$

gives the discrete number of events occurring within a given time interval. The parameter λ is the average number of events in the given time interval. This means that

$$\langle X \rangle = \lambda. \tag{2.56}$$

The variance is also equal to λ.

One can show (refer to statistics textbooks) that the Poisson distribution is a limiting case of the binomial distribution, Eq. 2.40 with $N \to \infty$, $p \to 0$ but with $Np = \lambda$, a constant.

2.8 One-dimensional random walk

This is an important problem first described by Lord Rayleigh in 1919. It is important because it finds applications in numerous scientific fields. We present a simple illustration of the problem.

Consider a drunkard walking on a narrow path, so that he can take either a step to the left or a step to the right. The path can be assumed one-dimensional and the size of each step always the same. Let p be the probability of a step to the right and q the probability of a step to the left, so that $p + q = 1$. If the drunkard is completely drunk, we can assume that $p = q = 1/2$. This is a one-dimensional random walk (Fig. 2.6).

Consider N steps, n_R to the right and n_L to the left, so that $n_R + n_L = N$. If m is the final location, so that $n_R - n_L = m$, what is the probability of any position m after N steps? Since any step outcome (right or left) is independent of the rest, the probability of a particular sequence of n_R and n_L independent steps is equal to the product of probabilities of each of the steps, i.e., $p^{n_R} q^{n_L}$. Because the specific sequence of steps is not important, we have to multiply this probability by the number of all distinct sequences of n_R and n_L steps to the right and left, respectively. Ultimately, the probability for the final position after N steps is

$$P(n_R, n_L) = \frac{N!}{n_R! n_L!} p^{n_R} q^{n_L}. \tag{2.57}$$

Example 2.15

If $N = 4$, the probability of $m = 0$, i.e., that $n_R = n_L = 2$, is

$$P(2, 2) = \frac{4!}{2!\,2!} 0.5^2\, 0.5^2. \tag{2.58}$$

Since $m = 2n_R - N$, Eq. 2.57 yields

$$P(n_R) = \frac{N!}{n_R!(N - n_R)!} p^{n_R} q^{N - n_R} \tag{2.59}$$

Figure 2.6 One-dimensional random walk example.

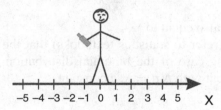

X

or

$$P(m) = \frac{N!}{\left[\frac{1}{2}(N+m)\right]! \left[\frac{1}{2}(N-m)\right]!} p^{1/2(N+m)} q^{1/2(N-m)}. \quad (2.60)$$

Consider the following question: what is the width of the distribution with respect to N for large N? In other words, where is the drunkard likely to be after N steps? Assuming $N \to \infty$, we can use Stirling's approximation and write

$$\lim_{N\to\infty} \ln(P(m)) = N \ln(N) + \frac{1}{2}\ln(N) - \frac{1}{2}\ln(2\pi) - N\ln(2) \quad (2.61)$$

or

$$\lim_{N\to\infty} \ln(P(m)) = \frac{1}{2}(N+m+1)\ln\left(\frac{N}{2}\right) \frac{1}{2}(N-m+1)\ln\left(\frac{N}{2}\right), \quad (2.62)$$

and finally

$$\lim_{N\to\infty} \ln(P(m)) = \frac{1}{2}(N+m+1)\ln\left[1+\left(\frac{m}{N}\right)\right]$$
$$-\frac{1}{2}(N+m+1)\ln\left[\left(1-\frac{m}{N}\right)\right]. \quad (2.63)$$

Remembering that

$$\ln(1+X) = X - \frac{1}{2}X^2 + \dots \quad (2.64)$$

we can expand the terms in these equations in a Taylor series, ignore higher than second order terms, and obtain

$$\lim_{N\to\infty} \ln(P(m)) = \frac{1}{2}\ln(N) - \frac{1}{2}\ln(2\pi) + \ln(2) - \frac{m^2}{N}, \quad (2.65)$$

or, simplifying

$$\lim_{N\to\infty} \ln(P(m)) = \ln\left(\frac{2}{\sqrt{2\pi N}}\right) - \frac{m^2}{2N}. \quad (2.66)$$

Exponentiating both sides yields

$$\lim_{N\to\infty} P(m) = \left(\frac{2}{\pi N}\right)^{1/2} \exp\left(-\frac{m^2}{2N}\right). \quad (2.67)$$

Therefore the binomial distribution becomes a Gaussian distribution for a large number of steps. This distribution peaks at $m = 0$.

An interesting question is related to the variance width and what happens to this width for very large N.

We can write

$$P(n_R) = \sqrt{\frac{2}{\pi N}} \exp\left[-\frac{(2n_R - N)}{2}\right]. \qquad (2.68)$$

The number of steps to the right, n_R, that maximizes the probability is determined as follows:

$$\frac{dP(n_R)}{dn_R} = 0 \qquad (2.69)$$

or

$$n_R = \frac{N}{2}. \qquad (2.70)$$

Again, the probability peaks at $N/2$ steps to the right and to the left. The variance is

$$\sigma^2 = N \qquad (2.71)$$

and the standard deviation

$$\sigma = \sqrt{N}. \qquad (2.72)$$

An important implication is that for very large numbers of steps, the distribution appears like a delta probability distribution. For example, if $N = 10^{24}$, the distribution is a bell-shaped curve approximately contained between $\frac{10^{24}}{2} \pm 10^{12}$. Although 10^{12} is an enormous number it is only 0.0000000001% of the average, which means that the distribution becomes apparently a delta function at $n_R = N/2$.

2.9 Law of large numbers

Consider N independent continuous variables $X_1, X_2, \ldots X_N$ with respective probability densities $\rho_1(X_1), \rho_2(X_2), \ldots \rho_N(X_N)$. We denote the joint probability with

$$\rho(X_1, X_2, \ldots X_N)dX_1\, dX_2 \ldots dX_N$$
$$= \rho_1(X_1)\rho_2(X_2)\ldots \rho_N(X_N)dX_1\, dX_2 \ldots dX_N. \qquad (2.73)$$

The joint probability is the one for

$$X_1 \in [X_1, X_1 + dX_1], \quad X_2 \in [X_2, X_2 + dX_2],$$
$$\ldots, \quad X_N \in [X_N, X_N + dX_N]. \qquad (2.74)$$

Let $\Delta X_1, \Delta X_2, \ldots \Delta X_N$ be the uncertainties of these random variables, with

$$(\Delta X_i)^2 = \langle (X_i - \langle X_i \rangle)^2 \rangle = \int (X_i - \langle X_i \rangle)^2 \rho_i(X_i) dX_i, \quad i = 1, 2 \ldots, N \tag{2.75}$$

where the average is

$$\langle X_i \rangle = \int X_i \rho_i(X_i) dX_i. \tag{2.76}$$

Define a new variable X as the average of the N original variables

$$X = \frac{X_1 + X_2 + \cdots + X_N}{N}. \tag{2.77}$$

What is the uncertainty of X? Let us start by computing the average $\langle X \rangle$ as follows:

$$\langle X \rangle = \int dX_1 \int dX_2 \ldots \int dX_N \frac{X_1 + X_2 + \cdots + X_N}{N}$$
$$\times \rho_1(X_1) \rho_2(X_2) \ldots \rho_N(X_N), \tag{2.78}$$

or, more concisely

$$\langle X \rangle = \frac{1}{N} \sum_{i=1}^{N} \int X_i \rho_i(X_i) dX_i, \tag{2.79}$$

which yields

$$\langle X \rangle = \frac{\langle X_1 \rangle + \langle X_2 \rangle + \cdots + \langle X_N \rangle}{N}. \tag{2.80}$$

For the uncertainty we can write

$$(\langle \Delta X \rangle)^2 = \langle (X - \langle X \rangle)^2 \rangle. \tag{2.81}$$

Expanding

$$(\langle \Delta X \rangle)^2$$
$$= \left\langle \left(\frac{(X_1 - \langle X_1 \rangle)^2 + (X_2 - \langle X_2 \rangle)^2 + \cdots + (X_N - \langle X_N \rangle)^2}{N} \right)^2 \right\rangle, \tag{2.82}$$

or

$$((\Delta X))^2 = \frac{1}{N^2} \int dX_1 \int dX_2 \ldots \int dX_N \sum_{i=1}^{N} \sum_{j=1}^{N}$$
$$\times (X_i - \langle X_i \rangle)^2 (X_j - \langle X_j \rangle)^2 \rho_i(X_i)\rho_j(X_j). \quad (2.83)$$

We can finally find that

$$((\Delta X))^2 = \frac{\sum_{i=1}^{N} ((\Delta X_i))^2}{N^2}. \quad (2.84)$$

This result is obtained with the help of the following relations:

$$\int \rho_i(X_i)dX_i = 1. \quad (2.85)$$

$$\int (X_i - \langle X_i \rangle)\rho_i(X_i)dX_i = 0, \quad (2.86)$$

and

$$\int (X_i - \langle X_i \rangle)^2 \rho_i(X_i)dX_i = ((\Delta X_i))^2. \quad (2.87)$$

Let us now define the mean squared uncertainty as

$$\alpha^2 = \frac{(\Delta X_1)^2 + (\Delta X_2)^2 + \cdots + (\Delta X_N)^2}{N^2}. \quad (2.88)$$

(If all variables had equal probabilities, then $\alpha = \Delta X_i$ for all i.)
The uncertainty in X is then

$$\Delta X = \frac{\alpha}{\sqrt{N}}. \quad (2.89)$$

This is an important result, which can be thought of as a generalization of the results found in the discussion of the random walk problem. According to Eq. 2.89 the average of N independent variables has an uncertainty that is only $\sqrt{N}$ as large as the uncertainties of the individual variables. This is generally called the law of large numbers.

2.10 Central limit theorem

A remarkable extension of the law of large numbers is the central limit theorem. This theorem states that the sum of a large number of independent random variables will be approximately normally distributed.

Importantly, the central limit theorem holds regardless of the probability distribution the random variables follow.

Specifically, if the independent random variables $X_1, X_2, \ldots, X_N$ have means $\langle X_1 \rangle, \langle X_2 \rangle, \ldots, \langle X_N \rangle$ and variances $\Delta X_1, \Delta X_2, \ldots, \Delta X_N$, respectively, then variable Y, calculated as

$$Y = \frac{\displaystyle\sum_{i=1}^{N} X_i - \sum_{i=1}^{N} \langle X_i \rangle}{\sqrt{\displaystyle\sum_{i=1}^{N} \Delta X_i}}, \tag{2.90}$$

is normally distributed, when N is large. The average $\langle Y \rangle$ is equal to zero and the variance is one.

In practice, for $N > 30$ the central limit theorem holds regardless of the underlying probability distribution function of the N independent variables (see T. Yamane's book, Further reading).

2.11 Further reading

1. Lord Rayleigh, On the problem of random vibrations and of random flights in one, two or three dimensions, *Phil. Mag.*, **37**, 321–347, (1919).
2. T. Yamane, *Statistics: An Introductory Analysis*, (New York: Harper and Row, 1967).
3. C. E. Weartherburn, *Mathematical Statistics*, (Cambridge: Cambridge University Press, 1962).
4. W. Mendenhall and T. Sincich, *Statistics for Engineering and the Sciences*, (Upper Saddle River, NJ: Prentice Hall, 2006).

2.12 Exercises

1. Plot the binomial distribution with the following parameters:
   a) $N = 10$, $p = 0.1$;
   b) $N = 10$, $p = 0.5$;
   c) $N = 1000$, $p = 0.1$.
   Comment on how well the binomial distribution approximates a Gaussian distribution.

2. Consider a system of N objects that can be in either of two states: up (U) or down (D). What is the number of possible arrangements? Assume that the probability of the particles being in either state is 0.5, and that their states are independent of one another. Also assume that all the possible arrangements have equal probabilities. What is the probability of finding a system within

a range 45–55% of the objects being in the up state if i) $N = 100$, ii) $N = 10\,000$, iii) an arbitrarily large number N. Consider the all-up and the all-down arrangements as the most ordered ones. What is the number of possible arrangements for the ordered states?

3. Consider the continuous probability distribution function

$$P(x) = \sqrt{\frac{\beta}{\pi}} \exp\left(-\beta x^2\right), \qquad (2.91)$$

where β is a constant. Calculate the average value of x, the average value of x^2, and the average value of $|x|$.

4. A device is made of two different parts. If the probability a part is defective is p_1 and p_2 for the two parts respectively, what is the probability that the device has zero, one, or two defective parts?

5. A carton contains 12 red balls and ten green balls. We randomly pick three balls and the first one is red. What is the probability that the other two are also red? If we pick one ball at a time, look at it and do not return it to the box, until we have found all ten green balls, what is the probability that the tenth green ball is found in i) the tenth trial, ii) the twentieth trial?

6. What is the relative error in using Stirling's approximation for $N = 100$? $N = 1000$?

7. A tire company knows that on average a tire fails after 100 000 miles. The number of tires failing is following a Poisson distribution. What is the probability that no more than one tire of a car fails in a 10 000 miles distance?

8. We place three coins in a box. One coin is two-headed. Another coin is two-tailed. Only the third coin is real, but they all look alike. Then we randomly select one coin from this box, and we flip it. If heads comes up, what is the probability that the other side of the coin is also heads?

9. The probability density of a continuous variable x is $\rho(x) = a(x + 3)$ for $2 \leq x \leq 8$. What is the value of a? What is the probability $P(3 < x < 5)$? What is the probability $P(X > 4)$?

10. The probability density of a random variable x is:

$$\rho(x) = \begin{cases} k, & \text{for } \theta \leq x \leq 2 \\ \theta, & \text{for } 0 \leq x \leq \theta \\ 0. & \text{otherwise} \end{cases}$$

What is the value of constant k as a function of θ? What is the average x?

11. Consider the birthdays of a group of N people. What is the probability that at least two people have the same birthday? What is the size of the probability space for this situation, i.e., what is the number of possible outcomes?

12. You are contemplating investing in real estate mutual funds. Get the file with the US Real Estate index from www.djindexes.com. Find the daily changes of the price returns and draw the probability density of daily changes. Find an analytical expression for the probability density to fit the data. Use this to find the average index change. Do the same, but instead of daily changes, use annual changes. If your financial advisor can guarantee an annual return of 1% over the market changes, and if you assume that the annual changes probability density will be the same in the future as in the past (this is not true and this is why financial predictions fail, but let's pretend we trust the financial advisor), and you invest $1000 what do you expect your return to be three years from now?

13. The probability density of a random variable x is:

$$\rho(x) = \begin{cases} cx^2, & \text{if } 1 \leq x \leq 2 \\ 0. & \text{otherwise} \end{cases}$$

   - What is the value of constant c?
   - Sketch the probability distribution function.
   - What is the probability that $x > 3/2$?

14. Show that $\binom{N}{n} = \binom{N-1}{n-1} + \binom{N-1}{n}$.

15. Suppose X has a normal distribution with mean 1 and variance 4. Find the value of the following: a) $P(X \leq 3)$ b) $P(|X| \leq 2)$.

16. A particle moves with an average speed of $450\,\text{ms}^{-1}$. It undergoes random elastic collisions with other particles after moving a distance equal to its mean free path, which is $100\,\text{nm}$. Determine the particle's root mean squared displacement after $10\,\text{s}$.

17. The velocity distribution of a gas is given by the Maxwell–Boltzmann distribution. What fraction of the molecules have a speed within 10% of the average speed?

3

Phase spaces, from classical to quantum mechanics, and back

Give me matter and motion, and I will construct the universe
René Descartes (1596–1650)

Consider a system with N interacting classical point-mass particles in constant motion. These are physical bodies whose dimensions can be assumed immaterial to their motion. Consider the particle positions $r_1, r_2, \ldots, r_N$, where $r_i = (x_i, y_i, z_i)$, for all i, in a Cartesian coordinate system.

Classical mechanics is the physicomathematical theory that describes the motion of classical bodies as a succession of their positions in time. Starting with the work of Isaac Newton, classical mechanics evolved in the eighteenth and nineteenth centuries with the work of Lagrange and Hamilton.

A major underpinning assumption in classical mechanics is that the positions of physical bodies can assume values that are arbitrarily close. This has proven to be a false assumption, and quantum mechanics has emerged as the proper physicomathematical theory that describes the motion of microscopic systems.

In this chapter, we present elements of classical mechanics and quantum mechanics that will prove useful in connecting microscopic and macroscopic thermodynamics. We introduce the concept of a system's phase space as a crucial one in statistical thermodynamics.

3.1 Classical mechanics

3.1.1 Newtonian mechanics

Newtonian mechanics is simply the application of Newton's second law for each particle i

$$m_i \ddot{r}_i = F_i, \tag{3.1}$$

where m_i is the mass of particle i, $\ddot{r}_i = d^2 r / dt^2$, and $\underline{F}_i$ is the force vector on particle i, exerted by the rest of the particles in the absence of any external force fields (Fig. 3.1). Then,

$$\underline{F}_i = F x_i \vec{i} + F y_i \vec{j} + F z_i \vec{k} = -\frac{\partial U}{\partial \underline{r}_i}, \tag{3.2}$$

where

$$\frac{\partial U}{\partial \underline{r}_i} = \left(\frac{\partial}{\partial x} \vec{i} + \frac{\partial}{\partial y} \vec{j} + \frac{\partial}{\partial z} \vec{k} \right) U. \tag{3.3}$$

In Eq. 3.2, $\vec{i}, \vec{j}, \vec{k}$ are the unit vectors of the Cartesian coordinate system, and $F x_i, F y_i, F z_i$ are the vector components of the force on particle i in the three directions x, y, z.

The potential energy U is a function of the positions of the particles

$$U = U(\underline{r}_1, \underline{r}_2, \ldots, \underline{r}_N). \tag{3.4}$$

For classical systems, the potential energy can be assumed to equal the sum of pairwise interactions between particles

$$U = \sum_{i>j} u_{ij}(|\underline{r}_i - \underline{r}_j|), \tag{3.5}$$

where u_{ij} is the interaction energy between any two particles i and j. This pairwise interaction energy is a function of the distance between two particles. Typically, u_{ij} can consist of dispersion and electrostatic interactions. We describe different types of interaction in Chapter 8.

The sum in Eq. 3.5 can be written equivalently, yet less compactly, as follows

$$\sum_{i>j} = \sum_{i=1}^{N-1} \sum_{j=i+1}^{N} = \frac{1}{2} \sum_{i \neq j}. \tag{3.6}$$

Figure 3.1 The force $\underline{F}_1$ on particle 1 is the vectorial sum of forces, $\underline{F}_{12}$ and $\underline{F}_{13}$, exerted on 1 by particles 2 and 3, respectively.

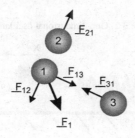

Newtonian mechanics is then a set of $3N$ second-order differential equations, which are coupled through the distance dependence of the potential function. One can show that these equations of motion conserve the total energy of the system of N particles. The total energy, E, is the potential plus the kinetic energies: $E = K + U$, as discussed in the Introduction.

Example 3.1
Consider a one-dimensional harmonic oscillator (Fig. 3.2). This is a body of mass m attached to a spring and moving horizontally in direction x on a frictionless surface. The potential energy is given by Hooke's law

$$U(x) = \frac{1}{2}kx^2, \tag{3.7}$$

where k is the spring constant and $x = 0$ is the equilibrium position with $U(x = 0) = 0$.

Newton's equation of motion in one dimension is

$$F = m\ddot{x}. \tag{3.8}$$

Using

$$F = -\frac{\partial U}{\partial x} \tag{3.9}$$

yields

$$\frac{d^2x}{dt^2} + \frac{kx}{m} = 0. \tag{3.10}$$

The solution of this linear, homogeneous differential equation is

$$x(t) = A \sin(\omega t) + B \cos(\omega t), \tag{3.11}$$

where A, B are constants and $\omega = \sqrt{k/m}$ is the vibrational frequency of the oscillator's motion.

Figure 3.2 One-dimensional oscillator.

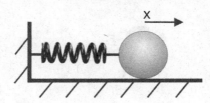

An alternative derivation can use the momentum of the oscillator,

$$p = m\dot{x}. \tag{3.12}$$

Then

$$\frac{dp}{dt} = -\frac{\partial U}{\partial x} \tag{3.13}$$

and

$$\frac{dx}{dt} = \frac{p}{m}. \tag{3.14}$$

Instead of one second-order, we end up with two first-order differential equations. The solution is, of course, the same as before.

Although conceptually satisfying, Newtonian mechanics is cumbersome to use when it is preferable to work with a coordinate system other than the Cartesian, such as the polar or the cylindrical coordinate systems. For example, it may be more convenient to replace the two Cartesian coordinates x, y of a particle moving in a plane, by two polar coordinates ρ, θ. Or, we may want to replace the six coordinates, x_1, y_1, z_1, x_2, y_2, z_2, of a pair of particles by the three coordinates X, Y, Z of the center of mass and the three coordinates x, y, z of the relative particle distance. Indeed, we will discuss molecular systems that are best described not by the Cartesian coordinates of all of their atoms, but by internal coordinates, such as bond lengths, bond angles and torsional angles. Casting Newton's equations of motion to an internal coordinate system is not practical in most cases.

The difficulties of Newtonian mechanics were discussed by Lagrange and Hamilton, among others. They proposed equivalent classical mechanical descriptions of matter, without the drawbacks of Newtonian mechanics, developing a general method for setting up equations of motion directly in terms of any convenient set of coordinates. Both methods begin by deriving the kinetic and potential energy of systems with N particles and derive equations of motion by manipulating these energy functions. In order to present Lagrangian and Hamiltonian mechanics, we first introduce the concept of generalized coordinates.

3.1.2 Generalized coordinates

Consider a complete set of $3N$ coordinates $\underline{q}_1, \underline{q}_2, \ldots, \underline{q}_N$ that precisely locate the positions of all N particles.

These are called generalized coordinates and can be related to the Cartesian coordinates through a coordinate transformation

$$\underline{r}_1 = \underline{r}_1(\underline{q}_1, \underline{q}_2, \ldots, \underline{q}_N)$$
$$\underline{r}_2 = \underline{r}_2(\underline{q}_1, \underline{q}_2, \ldots, \underline{q}_N)$$
$$\ldots$$
$$\underline{r}_N = \underline{r}_N(\underline{q}_1, \underline{q}_2, \ldots, \underline{q}_N). \tag{3.15}$$

Conversely, we can write

$$\underline{q}_1 = \underline{q}_1(\underline{r}_1, \underline{r}_2, \ldots, \underline{r}_N)$$
$$\underline{q}_2 = \underline{q}_2(\underline{r}_1, \underline{r}_2, \ldots, \underline{r}_N)$$
$$\ldots$$
$$\underline{q}_N = \underline{q}_N(\underline{r}_1, \underline{r}_2, \ldots, \underline{r}_N). \tag{3.16}$$

The mathematical condition that Eqs. 3.15 and 3.16 have a solution is that the Jacobian determinant of these equations be different from zero (see for example W. F. Osgood's book in Further reading).

Generalized velocities, $\underline{\dot{q}}_1, \underline{\dot{q}}_2, \ldots, \underline{\dot{q}}_N$ can be defined in a similar manner.

Example 3.2

Considering only one particle, a set of generalized coordinates describing the position of this particle can be the spherical coordinates $q = (R, \theta, \phi)$ (Fig. 3.3). These are related to the Cartesian coordinates through the following transformation

$$x = R \sin\theta \cos\phi$$
$$y = R \sin\theta \sin\phi$$
$$z = R \cos\theta. \tag{3.17}$$

Figure 3.3 Position of particle in spherical coordinates.

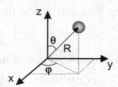

3.1.3 Lagrangian mechanics

Let us define the Lagrangian for a system of N particles as a function of the positions and velocities of the particles:

$$L(\underline{r}, \underline{\dot{r}}) = K(\underline{\dot{r}}) - U(\underline{r}), \tag{3.18}$$

where the shortcut notation is used for the positions and velocities of the particles

$$\underline{r} = \underline{r}_1, \underline{r}_2, \ldots, \underline{r}_N$$
$$\underline{\dot{r}} = \underline{\dot{r}}_1, \underline{\dot{r}}_2, \ldots, \underline{\dot{r}}_N. \tag{3.19}$$

Differentiating the Lagrangian with respect to the positions yields

$$\frac{\partial L}{\partial \underline{r}} = -\frac{\partial U}{\partial \underline{r}}. \tag{3.20}$$

Differentiating the Lagrangian with respect to the velocities results in

$$\frac{\partial L}{\partial \underline{\dot{r}}} = \frac{\partial K}{\partial \underline{\dot{r}}} = m\underline{\dot{r}}. \tag{3.21}$$

Differentiating Eq. 3.21 with respect to time results in

$$\frac{d}{dt}\left(\frac{\partial L}{\partial \underline{\dot{r}}}\right) = m\underline{\ddot{r}}. \tag{3.22}$$

According to Newton's second law of motion, the right-hand sides of Eq. 3.20 and Eq. 3.22 are equal. Consequently,

$$\frac{d}{dt}\left(\frac{\partial L}{\partial \underline{\dot{r}}}\right) = \frac{\partial L}{\partial \underline{r}}, \tag{3.23}$$

or, written more explicitly in terms of particle positions and velocities

$$\frac{d}{dt}\left(\frac{\partial L}{\partial \underline{\dot{r}}_i}\right) - \frac{\partial L}{\partial \underline{r}_i} = 0. \quad i = 1, 2, \ldots, N \tag{3.24}$$

These are the $3N$ equations of motion in Lagrangian mechanics. They generate the same solution for the motion of classical particles as Newton's equations of motion. The benefit of Lagrangian mechanics is that Eqs. 3.24 are invariant to coordinate transformations. This simply means that the functional form of the equations of motion remains unchanged for appropriate generalized coordinates

$$\frac{d}{dt}\left(\frac{\partial L}{\partial \dot{\underline{q}}_i}\right) - \frac{\partial L}{\partial \underline{q}_i} = 0. \quad i = 1, 2, \ldots, N \qquad (3.25)$$

To illustrate this property of Lagrange's equations of motion we start with Eq. 3.15. Differentiating with respect to time yields, in concise vectorial form

$$\frac{d\underline{r}_i}{dt} = \sum_{j=1}^{N} \frac{\partial \underline{r}_i}{\partial \underline{q}_j}\frac{d\underline{q}_j}{dt}, \quad i = 1, 2, \ldots, N \qquad (3.26)$$

or

$$\dot{\underline{r}}_i = \sum_{j=1}^{N} \frac{\partial \underline{r}_i}{\partial \underline{q}_j}\dot{\underline{q}}_j. \quad i = 1, 2, \ldots, N \qquad (3.27)$$

A note may be useful here regarding the notation. For the three coordinates (r_{i1}, r_{i2}, r_{i3}) of atom i, vector $\dfrac{d\underline{r}_i}{dt}$ is defined as

$$\begin{bmatrix} \dfrac{dr_{i1}}{dt} \\[2mm] \dfrac{dr_{i2}}{dt} \\[2mm] \dfrac{dr_{i3}}{dt} \end{bmatrix}, \qquad (3.28)$$

whereas the term $\dfrac{\partial \underline{r}_i}{\partial \underline{q}_j}$ is the Jacobian matrix,

$$\begin{bmatrix} \dfrac{\partial r_{i1}}{\partial q_{j1}} & \dfrac{\partial r_{i1}}{\partial q_{j2}} & \dfrac{\partial r_{i1}}{\partial q_{j3}} \\[3mm] \dfrac{\partial r_{i2}}{\partial q_{j1}} & \dfrac{\partial r_{i2}}{\partial q_{j2}} & \dfrac{\partial r_{i2}}{\partial q_{j3}} \\[3mm] \dfrac{\partial r_{i3}}{\partial q_{j1}} & \dfrac{\partial r_{i3}}{\partial q_{j2}} & \dfrac{\partial r_{i3}}{\partial q_{j3}} \end{bmatrix}. \qquad (3.29)$$

We can then also write

$$\frac{\partial \underline{r}_i}{\partial \dot{\underline{q}}_j} = \frac{\partial \underline{r}_i}{\partial \underline{q}_j}. \qquad (3.30)$$

The differential of the potential energy with respect to the generalized coordinates can be written as

$$-\frac{\partial U}{\partial \underline{q}_j} = -\sum_{i=1}^{N} \frac{\partial U}{\partial \underline{r}_i} \frac{\partial \underline{r}_i}{\partial \underline{q}_j}. \quad j = 1, 2, \ldots, N \tag{3.31}$$

Similarly, for the kinetic energy

$$\frac{\partial K}{\partial \underline{\dot{q}}_j} = -\sum_{i=1}^{N} \frac{\partial K}{\partial \underline{\dot{r}}_i} \frac{\partial \underline{\dot{r}}_i}{\partial \underline{\dot{q}}_j}. \quad j = 1, 2, \ldots, N \tag{3.32}$$

Multiplying both sides of Eq. 3.24 with $\partial \underline{r}_i / \partial \underline{q}_j$ yields

$$\frac{\partial \underline{r}_i}{\partial \underline{q}_j} \frac{d}{dt}\left(\frac{\partial K}{\partial \underline{\dot{r}}_i}\right) + \frac{\partial \underline{r}_i}{\partial \underline{q}_j} \frac{\partial U}{\partial \underline{r}_i} = 0. \quad i = 1, 2, \ldots, N \tag{3.33}$$

Summing up over all particles i yields

$$\sum_{i=1}^{N} \left(\frac{\partial \underline{r}_i}{\partial \underline{q}_j} \frac{d}{dt}\left(\frac{\partial K}{\partial \underline{\dot{r}}_i}\right)\right) + \frac{\partial U}{\partial \underline{q}_j} = 0. \tag{3.34}$$

Note that

$$\frac{\partial \underline{r}_i}{\partial \underline{q}_j} \frac{d}{dt} \frac{\partial K}{\partial \underline{\dot{r}}_i} = \frac{d}{dt}\left(\frac{\partial K}{\partial \underline{\dot{r}}_i} \frac{\partial \underline{r}_i}{\partial \underline{q}_j}\right) - \frac{\partial K}{\partial \underline{\dot{r}}_i} \frac{d}{dt} \frac{\partial \underline{r}_i}{\partial \underline{q}_j}. \tag{3.35}$$

Finally, note that

$$\frac{d}{dt} \frac{\partial \underline{r}_i}{\partial \underline{q}_j} = \sum_{k=1}^{N} \frac{\partial}{\partial \underline{q}_k}\left(\frac{\partial \underline{r}_i}{\partial \underline{q}_j}\right) \underline{\dot{q}}_k, \tag{3.36}$$

and because the differentiation order is immaterial

$$\frac{d}{dt} \frac{\partial \underline{r}_i}{\partial \underline{q}_j} = \sum_{k=1}^{N} \frac{\partial}{\partial \underline{q}_j}\left(\frac{\partial \underline{r}_i}{\partial \underline{q}_k}\right) \underline{\dot{q}}_k = \frac{\partial}{\partial \underline{q}_j} \sum_{k=1}^{N}\left(\frac{\partial \underline{r}_i}{\partial \underline{q}_k}\right) \underline{\dot{q}}_k, \tag{3.37}$$

or, finally

$$\frac{d}{dt} \frac{\partial \underline{r}_i}{\partial \underline{q}_j} = \frac{\partial \underline{\dot{r}}_i}{\partial \underline{q}_j}. \tag{3.38}$$

Introducing Eqs. 3.30, 3.34, and 3.38 in Eq. 3.35 we can write

$$\sum_{i=1}^{N}\left[\frac{d}{dt}\left(\frac{\partial K}{\partial \underline{\dot{r}}_i} \frac{\partial \underline{\dot{r}}_i}{\partial \underline{\dot{q}}_j}\right) - \frac{\partial K}{\partial \underline{\dot{r}}_i} \frac{\partial \underline{\dot{r}}_i}{\partial \underline{q}_j}\right] + \frac{\partial U}{\partial \underline{q}_j} = 0. \tag{3.39}$$

Consequently,

$$\frac{d}{dt}\left(\frac{\partial K}{\partial \dot{\underline{q}}_j}\right) - \frac{\partial K}{\partial \underline{q}_j} + \frac{\partial U}{\partial \underline{q}_j} = 0. \quad j = 1, 2, \ldots, N \qquad (3.40)$$

Lagrange's equations of motion are invariant to the coordinate transformation and Eq. 3.24 is equivalent to

$$\frac{d}{dt}\left(\frac{\partial K}{\partial \dot{\underline{q}}_j}\right) - \frac{\partial L}{\partial \underline{q}_j} = 0, \quad j = 1, 2, \ldots, N \qquad (3.41)$$

which, in turn, yields Eq. 3.25.

3.1.4 Hamiltonian mechanics

Changing the notation slightly, consider the generalized coordinates q_j of N particles, where $j = 1, 2, \ldots 3N$. This means that the first three coordinates, q_1, q_2, q_3, describe the position of particle 1, the second three, q_4, q_5, q_6, describe the position of particle 2, and so on. Note that these are different from the ones defined in Eq. 3.15. We will be using these two different notations interchangeably in the remainder of the book.

For each generalized coordinate, q_j, a generalized momentum, p_j, can be defined as

$$p_j = \frac{\partial L}{\partial \dot{q}_j} = m\dot{q}_j. \quad j = 1, 2, \ldots, 3N. \qquad (3.42)$$

Momentum p_j is said to be conjugate to coordinate q_j.

The Hamiltonian of a system with N particles of mass m is defined as the total energy

$$H(\underline{p}, \underline{q}) = \sum_{j=1}^{3N} \frac{p_j^2}{2m} + U(q_1, q_2, \ldots q_{3N}), \qquad (3.43)$$

where $\underline{p} = (p_1, p_2, \ldots, p_{3N})$, and $\underline{q} = (q_1, q_2, \ldots, q_{3N})$. Note that the generalized coordinates defined in Eq. 3.43 are numbered differently from the ones used in Eq. 3.25.

Differentiating Eq. 3.43 with respect to the coordinates and separately with respect to the momenta yields

$$\frac{\partial H(\underline{p}, \underline{q})}{\partial \underline{p}} = \frac{\underline{p}}{m} \qquad (3.44)$$

and

$$\frac{\partial H(\underline{p}, \underline{q})}{\partial \underline{q}} = \frac{\partial U(\underline{q})}{\partial \underline{q}}. \tag{3.45}$$

These can be written again as follows:

$$\frac{\partial H(\underline{p}, \underline{q})}{\partial \underline{p}} = \underline{\dot{q}} \tag{3.46}$$

and

$$\frac{\partial H(\underline{p}, \underline{q})}{\partial \underline{q}} = -\underline{\dot{p}}. \tag{3.47}$$

Finally, Hamilton's equations of motion are written as

$$\frac{\partial H}{\partial p_j} = \frac{dq_j}{dt},$$

$$\frac{\partial H}{\partial q_j} = -\frac{dp_j}{dt}, \tag{3.48}$$

for all $j = 1, 2, \ldots, 3N$.

Example 3.3

Consider again the one-dimensional oscillator. Its Hamiltonian is

$$H(p, x) = \frac{p^2}{2m} + \frac{1}{2}kx^2. \tag{3.49}$$

Differentiating the Hamiltonian with respect to the two degrees of freedom yields

$$\frac{\partial H}{\partial p} = \frac{p}{m} = \dot{x} \tag{3.50}$$

and

$$\frac{\partial H}{\partial x} = kx = -\dot{p}. \tag{3.51}$$

The second equation is the one we obtained applying Newtonian mechanics.

Hamilton's equations of motion also conserve the energy of the system, in the absence of external force fields. To prove this, we start by computing the total differential of $H(\underline{p}, \underline{q})$

$$dH(\underline{p}, \underline{q}) = \frac{\partial H(\underline{p}, \underline{q})}{\partial \underline{p}} d\underline{p} + \frac{\partial H(\underline{p}, \underline{q})}{\partial \underline{q}} d\underline{q}, \qquad (3.52)$$

or, expanding

$$dH(\underline{p}, \underline{q}) = \sum_{j=1}^{3N} \left(\frac{\partial H(\underline{p}, \underline{q})}{\partial p_j} dp_j + \frac{\partial H(\underline{p}, \underline{q})}{\partial q_j} dq_j \right). \qquad (3.53)$$

Differentiating with respect to time

$$\frac{dH}{dt} = \sum_{j=1}^{3N} \left(\frac{\partial H}{\partial p_j} \frac{dp_j}{dt} + \frac{\partial H}{\partial q_j} \frac{dq_j}{dt} \right). \qquad (3.54)$$

Using Hamilton's equations of motion yields

$$\frac{dH}{dt} = \sum_{j=1}^{3N} \left(-\frac{\partial H}{\partial p_j} \frac{\partial H}{\partial q_j} + \frac{\partial H}{\partial q_j} \frac{\partial H}{\partial p_j} \right) \qquad (3.55)$$

and

$$\frac{dH}{dt} = 0. \qquad (3.56)$$

Hamilton's equations of motion can precisely determine the microscopic state of any system, given initial conditions and the functional form of the potential energy, i.e., the pairwise interaction energy between particles as a function of their distance. Indeed, with an initial condition of values for $6N$ microscopic degrees of freedom, classical mechanics can be used to predict both the future and the past of any system of N particles, since all equations of motion are reversible, or, in other words, they are integrable in negative time. This remains in principle true even if N is equal to the number of all the particles in the universe. Of course there is no practical way to find a precise initial condition for large systems, let alone the universe, and certainly it is not feasible to integrate $6N$ equations of motion for large N.

Nonetheless, the philosophical conundrum that emerges from classical mechanics is that the cosmos moves from a precisely determined past to a precisely predetermined future. Marquis de Laplace perhaps said it best: "We may regard the present state of the universe as the effect of its past and the cause of its future. An intellect which at any given moment knew all the forces that animate nature and the mutual positions of the beings that compose it, if this intellect were vast enough to submit the data to analysis, could condense into a single formula the movement of the greatest bodies of the universe and the movement of the lightest atom; for such an intellect nothing could be uncertain and the future just like the past would be present before its eyes."

This is difficult to accept, since, dear reader, this very moment, during which you are reading these words, was predetermined at the moment of the Big Bang, 13.7 billion years ago. This is patently absurd. Before discussing how quantum mechanics has provided a theoretical framework that resolves these absurdities, we will present the concept of phase space, stressing that classical mechanics is still an eminently powerful and useful approximation of reality. Newtonian mechanics is the first self-contained system of physical causality humans developed and for this, using Einstein's words, "Newton deserves our deepest reverence."

3.2 Phase space

For a system with N particles, consider the $6N$-dimensional Cartesian space with $6N$ mutually orthogonal axes, each representing one of the $6N$ coordinates q and momenta p of the particles. We call this the phase space of the system, and denote it with Γ. This is simply a mathematical construct that allows us concisely to define the microscopic states of the system. Each point in phase space $\underline{X}$ is a microscopic state of the system:

$$\underline{X}(\underline{p}, \underline{q}) = (\underline{p}_1, \underline{p}_2, \ldots, \underline{p}_N, \underline{q}_1, \underline{q}_2, \ldots, \underline{q}_N) \tag{3.57}$$

or

$$\underline{X}(\underline{p}, \underline{q}) = (p_1, p_2, \ldots, p_{3N}, q_1, q_2, \ldots, q_{3N}). \tag{3.58}$$

We can now define a phase orbit or phase trajectory as the succession of microstates $\underline{X}(t)$, with momenta and coordinates $(\underline{p}(t), \underline{q}(t))$ being the solution of forward integration of Hamilton's equations of motion for a set of initial conditions $\underline{X}(0) = (\underline{p}(0), \underline{q}(0))$. The phase orbit can be simply considered as a geometrical interpretation of Hamilton's equations of motion: a system with N mass points, which move in three dimensions, is mathematically equivalent to a single point moving in $6N$ dimensions (Fig. 3.4).

As discussed in the previous section, knowledge of the microscopic state of a system $\underline{X}(t_1)$ at a particular time instance t_1 completely determines the microscopic states of the system for all times t_2 (both in the future, with $t_2 > t_1$, and in the past, with $t_2 < t_1$).

The following observations can be made regarding the phase space:

1. Distinct trajectories in phase space never cross. Proof: let $\underline{X}_A(t_2)$ and $\underline{X}_B(t_2)$ be two phase space points of two systems A and B characterized by the same Hamiltonian at the same time t_2. Assume that at this time t_2 they are different, i.e., $\underline{X}_A(t_2) \neq \underline{X}_B(t_2)$. Then the observation states that $\underline{X}_A(t_1) \neq \underline{X}_B(t_1)$, $\forall t_1$.

    If they did cross at time t_1, then by integrating the equations of motion for $t_1 - t_2$ yields $\underline{X}_A(t_2) = \underline{X}_B(t_2)$, which of course violates the initial premise. Therefore distinct trajectories never cross (Fig. 3.5).

2. Trajectories can never intersect themselves. Proof: the same logic that was used in the first observation can be used to prove this.

Example 3.4

Consider the harmonic oscillator. The microscopic state of a harmonic oscillator is completely defined by two degrees of freedom, its position q and its momentum p. The solutions of Hamilton's equations of motion for p and q are

$$q = \sqrt{\frac{2E}{m\omega^2}} \sin(\omega t + \alpha) \tag{3.59}$$

and

$$p = \sqrt{2mE} \cos(\omega t + \alpha), \tag{3.60}$$

Figure 3.4 A trajectory in phase space as a succession of points $\underline{X}(t)$ in time.

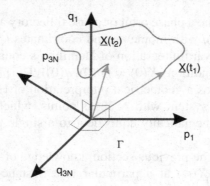

respectively, with α a constant of integration fixed by the initial conditions.

The phase space Γ of this oscillator is a two-dimensional plane with two orthogonal axes, p and q.

What does the phase orbit look like, describing the motion of the oscillator with a single point in phase space? In the absence of external influences, the energy of the oscillator will be constant. We can then write

$$H(p,q) = \frac{p^2}{2m} + \frac{kq^2}{2} = E \qquad (3.61)$$

or

$$\frac{p^2}{2mE} + \frac{kq^2}{2E} = 1. \qquad (3.62)$$

It is clear that the phase space trajectory for constant energy E is an ellipse on the pq plane (Fig. 3.6). It is, in other words, a one-dimensional object in a two-dimensional space. This could be expected, realizing that for constant energy the two variables of motion become dependent on one another, resulting in a single degree of freedom. The size of the

Figure 3.5 Impossible orbits in phase space.

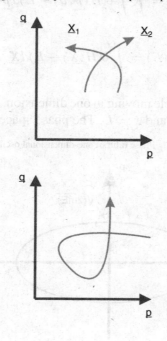

phase space that is available to the oscillator is simply equal to the perimeter of the ellipse.

3.2.1 Conservative systems

Consider a system of N particles evolving in time under the constraint of constant energy. The state vectors form a $(6N - 1)$-dimensional constant energy hypersurface of size

$$\Sigma(E) = \int_{\underline{p}_1} \int_{\underline{p}_2} \cdots \int_{\underline{p}_N} \int_{\underline{q}_1} \int_{\underline{q}_2} \cdots \int_{\underline{q}_N}$$
$$\times \delta(H(\underline{p}_1, \underline{p}_2, \ldots, \underline{p}_N, \underline{q}_1, \underline{q}_2, \ldots, \underline{q}_N) - E)d\underline{p}_1 d\underline{p}_2$$
$$\times \ldots d\underline{p}_N d\underline{q}_1 d\underline{q}_2 \ldots d\underline{q}_N, \tag{3.63}$$

where δ is the Dirac function, defined as

$$\delta(H(\underline{X}) - E) = \begin{cases} 1, & \text{if } H(\underline{X}) = E \\ 0. & \text{if } H(\underline{X}) \neq E \end{cases} \tag{3.64}$$

The Dirac δ function in Eq. 3.63 selects the points in phase space that correspond to microscopic states with energy E (Fig. 3.7).

The $6N$ integrals in Eq. 3.63 can be written more concisely as

$$\Sigma(E) = \int_{\underline{p}} \int_{\underline{q}} \delta(H(\underline{p}, \underline{q}) - E)d\underline{p}d\underline{q} \tag{3.65}$$

or

$$\Sigma(E) = \int_{\Gamma} \delta(H(\underline{X}) - E)d\underline{X}. \tag{3.66}$$

Example 3.5
Consider a single particle moving in one dimension, bouncing elastically on two walls at $x = 0$ and $x = L$. The phase space is two dimensional

Figure 3.6 Phase orbit of one-dimensional oscillator.

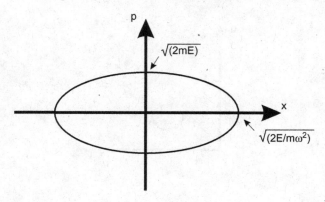

with two orthogonal axes, one for each of the two independent degrees of freedom, x and p_x. The phase orbit is one-dimensional, consisting of two lines at $\pm|p_x|$. The size of the available phase space is simply $\Sigma = 2L$ (Fig. 3.8).

Example 3.6

Consider a single particle with a constant energy $E = p^2/2m$, constrained on a two-dimensional surface (Fig. 3.9). The phase space is now four-dimensional. Note that two-dimensional projections are shown in Fig. 3.9.

3.3 Quantum mechanics

At this point, we need to bring quantum mechanics into the discussion. The ultimate goal of this section is to discuss how quantum mechanical principles discretize the phase space of systems.

Figure 3.7 Constant energy hypersurface Σ representing microscopic states of NVE system.

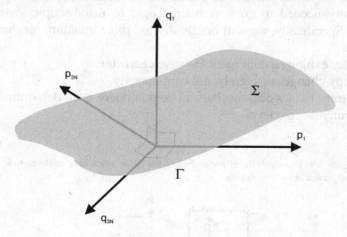

Figure 3.8 The phase space of a single particle constrained in one dimension.

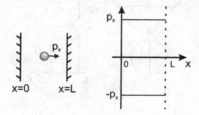

Quantum mechanics is a crowning achievement of human efforts to grasp and explain the natural world. It is often, however, void of experiential, physical insight, simply because its truths pertain to a world that is not familiar to human sensing capacities and experiences. Bohr, one of the founding fathers of quantum mechanics, said that "When it comes to atoms, language can be used only as in poetry. The poet, too, is not nearly so concerned with describing facts as with creating images and establishing mental connections." Simply put, our brains may be patently incapable of capturing the totality of quantum mechanical reality.

Certainly, experimental evidence and empirical phenomena are irrefutable and although even the best human minds, like Einstein's, find it difficult to accept quantum mechanical concepts, there is no question as to their truth. Thankfully, rich and intellectually satisfying insight can result from mathematical descriptions and interpretations of quantum mechanical phenomena.

In this chapter we rely on mathematical descriptions and interpretations to discuss quantum mechanical concepts only to the extent they can serve our purpose of presenting the statistical thermodynamics theory needed to connect microscopic to macroscopic states of matter. Specifically, we will briefly discuss three quantum mechanical dictums:

1. Matter exhibits a dual particle–wave character.
2. Energy changes discretely, not continuously.
3. Contrary to classical mechanical dogma, there is no determinism at the microscopic level.

Figure 3.9 Two-dimensional projections of the phase space of a single particle constrained in two dimensions.

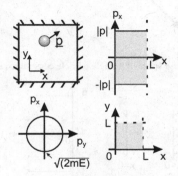

3.3.1 Particle–wave duality

In the 1800s the wave nature of light and the particulate nature of matter dominated physical explanations of the physical world. Newton's and Maxwell's equations were thought to capture all reality, conferring to it an inescapable, deterministic character. Then, at the end of the nineteenth century numerous experiments were conducted that were not reconciled with these prevailing classical mechanical notions.

In particular, black body radiation experiments by Max Planck in 1900 revealed unambiguously that energies of radiating particles are quantized, i.e., that they only attain discrete values which are integer multiples of the characteristic radiation frequency f:

$$E = nhf, \quad n = 0, 1, \dots \tag{3.67}$$

where $h = 6.624 \times 10^{-27}$ erg is Planck's constant.

Planck's careful experiments demonstrated that particles go from a higher to a lower energy by emitting a quantum (packet) of light radiation. Einstein likened these light packets to small particles he called photons.

Separately, the photoelectric effect and the Compton effect experiments unambiguously demonstrated that light photons behave like particles do. And around the same time the Davisson–Germer experiments showed how electrons exhibit wave properties.

The magnificent implication of these experimental observations is that all matter exhibits both a particulate and a wave character, and de Broglie was the first to quantify the connection of the two characters of matter.

For a light photon it was known that

$$E = hv, \tag{3.68}$$

where v is the frequency of light.

With special relativity theory, Einstein showed that for any moving particle the energy

$$E = mc^2, \tag{3.69}$$

where m is the relativistic mass of the particle and c is the speed of light.

de Broglie simply equated the two energies of a particle and a photon,

$$hv = mc^2. \tag{3.70}$$

Using the relationship between the frequency v and the wavelength λ of light,

$$v = c/\lambda, \tag{3.71}$$

yields the wavelength of any particle of mass m,

$$\lambda = \frac{h}{mc}. \tag{3.72}$$

de Broglie propounded the idea that this relationship between the wavelength and the momentum can be extended to all particles with non-zero rest mass m and velocity u, arguing that all such particles have a wavelength λ, even when u is significantly smaller than the speed of light, that is

$$\lambda = \frac{h}{mu} \tag{3.73}$$

or

$$\lambda = \frac{h}{p}. \tag{3.74}$$

Of course, likening the character of particles to that of waves is only an analogy. So what does it mean that particles behave like waves and have an associated wavelength? And why does the energy of particles and light change only discontinuously?

Schrödinger derived an equation that affords elegant mathematical answers to these questions. Let us derive Schrödinger's equation, which describes the motion of particles as a probabilistic phenomenon and show how solving this equation is only possible when the energy is quantized.

We use an approximate proof that illustrates the connection between the particulate and wave nature of matter. We start with a classical wave and a classical particle and we use de Broglie's relations to connect the two. We then discuss what this connection means for quantum mechanical systems.

We begin with the classical, one-dimensional wave equation:

$$\frac{\partial^2 \phi}{\partial x^2} = \frac{1}{u^2} \frac{\partial^2 \phi}{\partial t^2}, \tag{3.75}$$

where $\phi(x, t)$ is the displacement, u is the velocity of propagation, and t is time.

We can define a new function $\psi(x)$, which depends only on the position, and write

$$\phi(x, t) = \psi(x)e^{2\pi i \nu t}, \tag{3.76}$$

where $\nu = u/\lambda$ is the frequency of the wave motion.

By definition

$$e^{2\pi i \nu t} = \cos(2\pi \nu t) + i\sin(2\pi \nu t). \tag{3.77}$$

Separation of variables leads to an ordinary differential equation

$$\frac{d^2\psi}{dx^2} + \frac{4\pi^2\nu^2}{u^2}\psi = 0. \tag{3.78}$$

This is the time-independent, or standing, wave equation.

Consider now a classical particle. The energy of the particle is

$$E = \frac{p^2}{2m} + U, \tag{3.79}$$

where U is the external force field potential the particle is experiencing.

Solving for the momentum yields

$$p = \sqrt{2m(E - U)}. \tag{3.80}$$

Using de Broglie's relationship between the wavelength and the momentum further yields

$$\lambda = \frac{h}{p} = \frac{h}{\sqrt{2m(E - U)}}. \tag{3.81}$$

The vibrational frequency of the wave is defined as

$$\nu = \frac{u}{\lambda}. \tag{3.82}$$

Combining Eq. 3.81 and Eq. 3.82 yields

$$\nu^2 = \frac{2mu^2(E - U)}{h^2}. \tag{3.83}$$

Equation 3.78 now becomes

$$\frac{d^2\psi}{dx^2} + \frac{8\pi^2 m}{h^2}(E - U)\psi = 0. \tag{3.84}$$

This is the one-dimensional, steady state Schrödinger equation.

We can generalize to three dimensions using the Laplacian operator

$$\nabla^2 = \frac{\partial^2}{\partial x^2} + \frac{\partial^2}{\partial y^2} + \frac{\partial^2}{\partial z^2} \tag{3.85}$$

to write

$$\nabla^2 \psi + \frac{8\pi^2 m}{h^2}(E - U)\psi = 0. \tag{3.86}$$

We can write Eq. 3.86 more concisely with the help of the Hamiltonian operator, which is defined as

$$H \equiv -\frac{\hbar^2}{2m}\nabla^2 + U, \tag{3.87}$$

where $\hbar = h/(2\pi)$.

The three-dimensional, steady state Schrödinger equation then becomes

$$H\psi = E\psi. \tag{3.88}$$

Schrödinger demonstrated that this equation can describe the dual wave–particle character of nature and explain how matter only attains discrete energy values.

The function ψ is called the wavefunction and describes the probability of finding a particle in a specific position. The functional form of the wavefunction resembles the functional form of classical waves. This is the reason for its name, and we should stress again that this is only a mathematical metaphor.

We can now present examples to find ψ and relate it to probabilities. Before that, we will discuss how the Schrödinger equation can be solved only for discrete values of energy.

Equation 3.88 belongs to a class of differential equations in which an operator acts on a function (the eigenfunction) and returns the function multiplied by a scalar (eigenvalue). Generally,

$$O(y) = ry, \tag{3.89}$$

where $O()$ is the operator, y is the eigenfunction, and r is the eigenvalue.

As an example, consider the simple differential equation

$$\frac{dy(x)}{dx} = ry(x). \tag{3.90}$$

Here the operator is d/dx and the eigenfunction is $y(x)$.

Equation 3.90 has the following solution:

$$\begin{cases} \text{eigenfunction,} & y = e^{ax}, \\ \text{eigenvalue,} & r = a. \end{cases} \tag{3.91}$$

Another example is the equation

$$\frac{d^2 y(x)}{dx^2} = r y(x), \tag{3.92}$$

with the following solution:

$$\begin{cases} \text{eigenfunction,} & y = A\cos(kx) + B\sin(kx), \\ \text{eigenvalue,} & r = f(A, B, k), \end{cases} \tag{3.93}$$

where $f(A, B, k)$ is a function of the constants A, B, and k.

In Schrödinger's equation, H is the operator, ψ is the eigenfunction, and E is the eigenvalue. It turns out that solutions exist only for certain discrete values of energy eigenvalues. A continuously changing energy will often result in a mathematically impossible solution of Schrödinger's equation.

Example 3.7
Consider the one-dimensional quantum harmonic oscillator, where a particle of mass m is moving under the influence of a harmonic potential

$$U(x) = \frac{1}{2}kx^2 = \frac{1}{2}m\omega^2 x^2, \tag{3.94}$$

where the angular frequency

$$\omega = \sqrt{\frac{k}{m}}. \tag{3.95}$$

Schrödinger's equation for the harmonic oscillator is

$$-\frac{\hbar^2}{2m}\frac{d^2\psi(x)}{dx^2} + \frac{1}{2}m\omega^2 x^2 \psi(x) = E\psi(x), \tag{3.96}$$

where $\psi(x)$ describes the probability of finding the particle at a particular interval $[x, x + dx]$.

A general solution of Eq. 3.96 exists only for a sequence of evenly spaced energy levels

$$E_n = \left(n + \frac{1}{2}\right)\hbar\omega, \quad \text{with } n = 0, 1, 2, \ldots \tag{3.97}$$

characterized by the quantum number n.

For each eigenvalue E_n a specific eigenfunction $\psi_n(x)$ exists. For example, the ground state of the system with $n = 0$ has an energy $E_0 = \hbar\omega$ and a wavefunction

$$\psi_0(x) = \left[\left(\frac{m\omega}{\pi\hbar} \right)^{\frac{1}{4}} \exp\left(-\frac{m\omega}{2\hbar}x^2 \right) \right]. \tag{3.98}$$

The potential energy, allowed energy levels, and wavefunctions for a harmonic oscillator are shown in Fig. 3.10 for the first few energy values.

A value of energy for the particle that does not satisfy Eq. 3.98 results in no solution for Schrödinger's equation. We see then how the discreteness of energy levels attainable by matter emerges naturally with the help of mathematics. It is interesting to note that the transition from one energy eigenvalue to the next is equal to $\hbar\omega$. Therefore as the mass of the particle increases, the transition energy decreases. For large, macroscopic size bodies the quantum transition energies are so small that for all practical purposes the energy can be considered a continuous variable.

Example 3.8
Consider a particle moving in a one-dimensional square well, which can be represented with the following potential energy:

$$U(x) = \begin{cases} \infty, & x < 0 \\ 0, & 0 \le x \le L \\ \infty. & x > L \end{cases} \tag{3.99}$$

Figure 3.10 Potential energy, U, allowed energy levels, E_n, and wavefunctions of a harmonic oscillator.

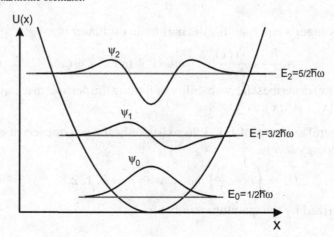

Schrödinger's equation in the region $0 \leq x \leq L$ is

$$\frac{\partial^2 \psi(x)}{\partial x^2} + \frac{8\pi^2 m}{h^2} E\psi = 0. \tag{3.100}$$

Outside the well the wavefunction vanishes, because of the infinite potential energy. We then have the boundary conditions

$$\psi(0) = \psi(L) = 0. \tag{3.101}$$

Solving Eq. 3.100 yields for the energy levels

$$E_n = \frac{h^2 n^2}{8mL^2}, \quad n = 1, 2, 3, \ldots \tag{3.102}$$

The corresponding wavefunctions are

$$\psi_n(x) = \sqrt{\frac{2}{L}} \sin\left(\frac{n\pi}{L}x\right). \tag{3.103}$$

The probability density of finding the particle at a position in $[0, L]$ is calculated from the wavefunction as follows:

$$\rho_n(x) = \psi_n^*(x)\psi_n(x), \tag{3.104}$$

where $\psi_n^*(x)$ is the complex conjugate of $\psi_n(x)$.
Consequently,

$$\rho_n(x) = \frac{2}{L} \sin^2\left(\frac{n\pi}{L}x\right). \tag{3.105}$$

Example 3.9
Consider a 5 kg mass attached to a spring with a spring constant $k = 400\,\mathrm{Nm}^{-1}$ undergoing simple harmonic motion with amplitude $A = 10\,\mathrm{cm}$. Assume that the energy this mass can attain is quantized according to Eq. 3.97. What is the quantum number n?
The potential energy is

$$E = \frac{1}{2}kA^2 = 2\,\text{joule}. \tag{3.106}$$

The frequency of oscillations is

$$f = \frac{\omega}{2\pi} = \frac{1}{2\pi}\sqrt{\frac{k}{m}} = \frac{1}{2\pi}\sqrt{\frac{400}{5}} = 1.42\,\text{Hz}. \tag{3.107}$$

Therefore

$$n = \frac{E}{hf} - \frac{1}{2} = 2.12 \times 10^{33}. \tag{3.108}$$

If the oscillator lowers its energy by a quantum then the change in the energy is

$$\Delta E = (n+1)hf - nhf = 9.4 \times 10^{-34} \text{ J}. \tag{3.109}$$

This is an astonishingly small change, which leads to the comfortable conclusion that quantization effects can be neglected for large, macroscopic bodies. Indeed, there can be no experiment that can detect this small an energy discontinuity. The implication is that for all practical purposes the energy is a continuous function for these large bodies.

The brilliance of Schrödinger's equation lies with its ability to capture the reality of microscopic systems in a way that eluded classical mechanical formalisms. For example, the motion of the single electron in the hydrogen atom cannot be described with Hamilton's equations of motion, assuming a simple Coulombic interaction between the electron and the proton. This hampered the efforts of scientists to develop a model of the atom that could explain experimental observations.

In a monumental achievement, Wolfgang Ernst Pauli used Schrödinger's equation to solve for the motion of electrons in atoms and developed his now famous exclusion principle, which forms the foundation of the structure of the periodic table of elements, and indeed of modern chemistry.

Let us consider the electron of the hydrogen atom (Fig. 3.11). Schrödinger's equation can be best written for the wavefunction as a

Figure 3.11 Simple model of the hydrogen atom.

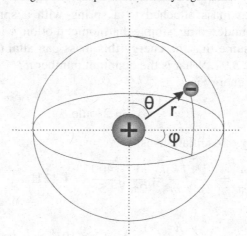

function in spherical coordinates

$$\psi = \psi(r, \theta, \phi).$$ (3.110)

Schrödinger's equation is then

$$-\frac{\hbar^2}{2\mu} \frac{1}{r^2 \sin\theta} \left[\sin\theta \frac{\partial}{\partial r} \left(r^2 \frac{\partial\psi}{\partial r} \right) + \frac{\partial}{\partial\theta} \left(\sin\theta \frac{\partial\psi}{\partial\theta} \right) + \frac{1}{\sin\theta} \frac{\partial^2\psi}{\partial\varphi^2} \right]$$
$$+ U(r)\psi(r, \theta, \phi) = E\psi(r, \theta, \phi),$$ (3.111)

where

$$\mu = \frac{m_e m_p}{m_e + m_p}$$ (3.112)

is the reduced mass of the electron-proton system, and

$$U(r) = -\frac{e^2}{4\pi\,\varepsilon_o r}$$ (3.113)

is the Coulomb interaction potential, with ε_o the dielectric constant of vacuum.

Equation 3.111 can be solved by separation of variables so that

$$\psi(r, \theta, \phi) = R(r)T(\theta)F(\phi).$$ (3.114)

The energy eigenvalues are found to be

$$E_n = -\frac{m_e e^4}{8\varepsilon_o h^2} \frac{1}{n^2} = -\frac{13.6\,\text{eV}}{n^2}.$$ (3.115)

In particular, solutions for $R(r)$ exist only if $n = 1, 2, 3, \ldots$. In turn, for each $R(r)$ solution related to a particular n, solutions exist for $T(\theta)$, if and only if another constant l, which arises in the solution, attains discrete values $l = 0, 1, 2, \ldots, n - 1$. Finally for each solution of $T(\theta)$, solutions exist for $F(\phi)$ when a third solution constant, m_l, attains the following discrete values $m_l = -l, (-l + 1), \ldots, (l - 1), l$. Here, n is called the principal quantum number, l is the orbital quantum number, and m_l is the magnetic quantum number. A fourth quantum number, m_s, is also necessary to describe the spin of electrons. The spin quantum number can take only one of two values, $+1/2$ or $-1/2$. Pauli extended this analysis to larger atoms and postulated that no two electrons can occupy a particular state described by the solution of Schrödinger's equation. In other words, no two electrons may have the same quantum numbers. This is Pauli's exclusion principle, which led to the orbital theory and the model of atomic structures used to this day.

The four quantum numbers completely describe the structure of electrons in atoms. More precisely, electrons with the same value of the principal quantum number will occupy the same electron shell, and electrons with the same value of the orbital quantum number will occupy the same subshell. The various shells and subshells, also known as orbitals, are designated by letters according to the following schemes:

value of n: 1 2 3 4 ...
symbol: K L M N ...

and

value of l: 0 1 2 3 ...
symbol : s p d f ...

For $n = 1$, l can only have one value, $l = 0$. Shell K then has only one subshell, s. For $n = 2$, l can have two values, $l = 0$ and $l = 1$. Shell L then has two subshells, s and p. Similarly, shell M has three subshells, s, p, and d and so on.

For each value of l, the corresponding subshell can accommodate up to $2(2l + 1)$ electrons, with different magnetic and spin quantum numbers. Thus the K shell can contain at most two electrons, the L shell can contain at most $2 + 6 = 8$ electrons, the M shell can contain at most $2 + 6 + 10 = 18$ electrons and so on. In general, the shell corresponding to n can contain at most $2n^2$ electrons.

For an atom with an atomic number of Z, electrons will occupy different orbitals, filling the lowest energy orbitals first. The order of filling is then 1s, 2s, 2p, 3s, 3p, 4s, 3d, 4p, 5s, 4d, 5p, 6s, and so on.

The electronic configurations of the first few elements derived using Pauli's exclusion principle are shown in Table 3.1.

3.3.2 Heisenberg's uncertainty principle

We can write Schrödinger's equation for a particle system that is not at steady state to describe the time-dependent wavefunction $\Psi(x, t)$ as follows:

$$i\frac{\partial}{\partial t}\Psi(x, t) = -\frac{\hbar^2}{2m}\frac{\partial^2}{\partial x^2}\Psi(x, t) + U(x, t)\Psi(x, t). \tag{3.116}$$

The derivation of this equation can be found in quantum mechanics textbooks (see Further reading). We are interested in the solution for

Table 3.1 *Electronic configurations of the first few elements in the periodic table.*

Z	Element	$n = 1$, $l = 0$	$n = 2$, $l = 0$	$n = 2$, $l = 1$	$n = 3$, $l = 0$	$n = 3$, $l = 1$	$n = 3$, $l = 2$	Electron configuration
1	Hydrogen	1						$1s^1$
2	Helium	2						$1s^2$
3	Lithium	2	1					$1s^2 2s^1$
4	Beryllium	2	2					$1s^2 2s^2$
5	Boron	2	2	1				$1s^2 2s^2 2p^1$
6	Carbon	2	2	2				$1s^2 2s^2 2p^2$
7	Nitrogen	2	2	3				$1s^2 2s^2 2p^3$
8	Oxygen	2	2	4				$1s^2 2s^2 2p^4$
9	Fluorine	2	2	5				$1s^2 2s^2 2p^5$
10	Neon	2	2	6				$1s^2 2s^2 2p^6$
11	Sodium	2	2	6	1			$1s^2 2s^2 2p^6 3s^1$
12	Magnesium	2	2	6	2			$1s^2 2s^2 2p^6 3s^2$
13	Aluminum	2	2	6	2	1		$1s^2 2s^2 2p^6 3s^2 3p^1$
14	Silicon	2	2	6	2	2		$1s^2 2s^2 2p^6 3s^2 3p^2$

this wavefunction which can be found to have the form

$$\Psi(x, t) = A \exp\left(2\pi i \left(\frac{x}{\lambda} - vt\right)\right), \tag{3.117}$$

so that if Ψ^* is the complex conjugate of Ψ then

$\Psi\Psi^* dx =$ the probability of finding the particle in the interval $[x, x + dx]$ at time t. $\tag{3.118}$

The constant A in Eq. 3.17 is a normalizing constant so that

$$\int_{-\infty}^{+\infty} \Psi\Psi^* dx = 1. \tag{3.119}$$

The average position of the particle is

$$<x> = \int_{-\infty}^{+\infty} x\Psi\Psi^* dx. \tag{3.120}$$

The uncertainty in the position can be expressed as follows:

$$(\Delta x) = \sqrt{(\Delta x)^2} = \sqrt{(<x^2> - <x>^2)}. \tag{3.121}$$

According to Schrödinger's equation, particle positions can be determined only with the help of probability distributions.

It turns out that particle velocities, or momenta, can also be determined only with the help of probability distributions. One can write an equivalent Schrödinger's equation for the momentum of a particle p. One can solve this equation and determine the momentum eigenfunction $\Phi(p)$ for different energy eigenvalues. With the help of the momentum eigenfunction, one can finally calculate the average momentum and the uncertainty in the momentum, with

$$< p >= \int_{-\infty}^{+\infty} p\Phi\Phi^*dq \qquad (3.122)$$

and

$$(\Delta p) = \sqrt{(\Delta p)^2} = \sqrt{(< p^2 > - < p >^2)}. \qquad (3.123)$$

Because of the nature of the eigenfunction solutions, Heisenberg determined that the uncertainties in the position and the momentum of a particle are not independent. The proof is beyond the scope of this book, but ultimately Heisenberg proved that in general

$$\Delta x \Delta p \geq h/4\pi. \qquad (3.124)$$

This is the celebrated Heisenberg's uncertainty principle, which states that the position and the momentum of a particle cannot be both determined with precision higher than h.

3.4 From quantum mechanical to classical mechanical phase spaces

An important implication of Heisenberg's uncertainty principle is that a microscopic state of a particle cannot be determined in phase space with resolution higher than $\Delta x \Delta p$. More accurately, there is only one microscopic state per $\Delta x \Delta p$ area in phase space for a single particle moving in one direction.

To be more precise, let us examine the example of a particle confined in a one-dimensional box of length L.

Earlier in this chapter, we found that the energy levels are

$$E_n = \frac{h^2 n^2}{8mL^2}, \quad n = 1, 2, 3, \ldots \qquad (3.125)$$

The energy of the system is solely kinetic. The particle momentum is then

$$p_n = \pm\frac{hn}{2L}, \quad n = 1, 2, 3, \ldots \qquad (3.126)$$

The phase space of this simple system is two-dimensional. A classical orbit is a pair of lines parallel to the x axis at $+p$ and $-p$, as discussed earlier in this chapter.

According to quantum mechanics, the values of the momentum p cannot change continuously. Instead, two phase orbits corresponding to two successive quantum numbers, r and $r + 1$, differ by

$$\frac{h(r + 1)}{2L} - \frac{hr}{2L}. \qquad (3.127)$$

The area between these two phase orbits, shown in Fig. 3.12, represents the highest possible resolution in phase space. The size of this area in phase space is simply

$$2L \left[\frac{h(r + 1)}{2L} - \frac{hr}{2L} \right] = h. \qquad (3.128)$$

Therefore, the phase space of a single particle moving in one dimension can be divided into cells of size h. The boundaries of the cells are classical orbits corresponding to successive quantum states.

This result can be generalized to any type of motion (e.g., rotational or vibrational) with the area of the phase space between classical orbits that correspond to successive quantum states being always precisely h. This means that for a phase space of volume Σ, the number of discrete accessible states will be Σ/h.

This result can also be generalized to a system of N particles. Consequently, we can infer that the phase space for a system of N particles in

Figure 3.12 Discrete nature of phase space.

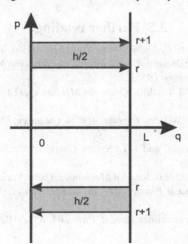

volume V with energy E is not a continuum with an infinite number of microscopic states. Instead, it contains a very large but finite number of microscopic states, each confined in an area of size h^{3N}. Consequently, if the volume of the phase space is Σ (the phase space volume Σ and the system volume V should not be confused), the number of discrete microscopic states is Σ/h^{3N}.

We see in the following chapters that if the N particles are not distinguishable, the actual number of unique microscopic states is $\Sigma/(N!h^{3N})$.

3.4.1 Born–Oppenheimer approximation

We should note that in the statistical mechanical treatment of matter examined in this book, the assumption will be implicitly made that the Born–Oppenheimer approximation is valid. Born and Oppenheimer (see Further reading for reference) used the fact that the mass of every atomic nucleus is several thousand times greater than the mass of an electron and determined that at temperatures less than 10 000 K the nucleus of atoms is at the ground energy state. The Hamiltonian term of nuclear interactions is then constant and does not have to be included in the calculation of the system's overall Hamiltonian. Electronic degrees of freedom will also be assumed to be negligible for ideal gases. For higher density systems, the interactions between electron clouds of different atoms will be taken into account. Even then though, the Born–Oppenheimer approximation will be considered valid.

3.5 Further reading

1. W. F. Osgood, *Advanced Calculus*, (New York: Macmillan, 1937).
2. J. von Neumann, *Mathematical Foundations of Quantum Mechanics*, (Princeton: Princeton University Press, 1955).
3. L. D. Landau and E. M. Lifshitz, *Quantum Mechanics*, (London: Pergamon Press, 1958).
4. L. Pauling and E. B. Wilson, *Introduction to Quantum Mechanics*, (New York: McGraw-Hill, 1935).
5. H. Goldstein, C. P. Poole, and J. L. Safko, *Classical Mechanics*, (San Francisco: Addison Wesley, 2002).
6. H. C. Corben and P. Stehle, *Classical Mechanics*, (New York: Dover, 1994).
7. U. Fano and L. Fano, *Basic Physics of Atoms and Molecules*, (New York: John Wiley, 1959).
8. M. Born and J. R. Oppenheimer, *Ann. d. Phys.*, **84**, 457, (1927).

3.6 Exercises

1. A particle moves with kinetic energy K_1 in a region where its potential energy has a constant value U_1. After crossing a certain plane, its potential energy changes discontinuously to a new constant value, U_2. The path of the particle makes an angle ϕ_1 with the normal to the plane before crossing and an angle ϕ_2 after crossing. This is the mechanical analogue of a ray of light crossing the boundary between two media of differing refractive indices. Determine the Lagrangian of the particle. Then derive the analog of Snell's Law, i.e., find the ratio $\sin\phi_1/\sin\phi_2$.

2. Prove that the kinetic energy of a system of N particles is expressed in terms of generalized coordinates as follows:

$$K(\underline{q}) = \frac{1}{2}\sum_{i=1}^{3N}\sum_{j=1}^{3N}\sum_{k=1}^{3N} m_i \frac{\partial r_i}{\partial q_j}\frac{\partial r_i}{\partial q_k}\dot{q}_j\dot{q}_k.$$

3. Show that

$$H = \sum_{j=1}^{3N}\dot{q}_j\frac{\partial L}{\partial q_j} - L.$$

4. Write the Hamiltonian of a particle in terms of a) Cartesian coordinates and momenta, and b) spherical coordinates and their conjugate momenta. Show that Hamilton's equations of motion are invariant to the coordinate transformation from Cartesian to polar coordinates.

5. Consider a system with the following Lagrangian:

$$L = \frac{1}{2}(\dot{q}_1^2 + \dot{q}_1\dot{q}_2 + \dot{q}_2^2).$$

Determine the Hamiltonian. Derive the equations of motion.

6. Consider a system of volume V with N point mass particles. Assume the system can exchange energy with its environment. In this case, the Hamiltonian and the Lagrangian will be explicit functions of time, i.e.

$$H = H(p, q, t)$$

and

$$L = L(\dot{q}, q, t).$$

Prove that if the energy of the system changes, then

$$\frac{\partial H}{\partial t} = -\frac{\partial L}{\partial t}.$$

7. Consider a rigid rotor consisting of two point mass particles of mass m joined together at a distance r. Express the kinetic energy of the rigid rotor in terms of

Cartesian coordinates of the two particles. Express the kinetic energy in terms of the following generalized coordinates: X, Y, Z are the coordinates of the rotor center of mass; r is the distance between the two particles; θ and ϕ are the polar angles of the line joining the two masses, referred to the Cartesian coordinates.

8. Consider two particles with masses m_1 and m_2, at positions $\underline{r}_1$ and $\underline{r}_2$ with velocities $\underline{v}_1$ and $\underline{v}_2$. Let $\underline{F}_1^{ex}$ and $\underline{F}_2^{ex}$ be external forces acting on each particle. Let $\underline{F}_1^{in}$ and $\underline{F}_2^{in}$ be forces exerted by each particle on the other ($\underline{F}_1^{in} = -\underline{F}_2^{in}$).

   Introduce a new set of coordinates, $\underline{R}$ and $\underline{r}$ for the center of mass of the two particle system and for their relative distance. Determine the conjugate momenta and write the kinetic energy in terms of these new momenta. Determine the equations of motion, for both the old and the new set of coordinates.

9. Consider two particles in an ideal gas phase. The particles are constrained to move in a single dimension, which is bounded by two walls at $x = 0$ and $x = L$. How many dimensions does the phase space of this simple system have? What is the size of the available phase space, Σ? What is the number of microscopic states, Ω?

10. Consider a single particle with a constant energy $E = p^2/2m$, constrained on a two-dimensional surface. What is the size, Σ, of the available classical mechanical phase space? What is the number of microscopic states, Ω?

11. Consider a classical, one-dimensional harmonic oscillator under the influence of a viscous force $-\gamma \dot{x}$. Assume that the initial position and velocity are $x(0)$ and $\dot{x}$, respectively. Draw the phase orbit of this oscillator.

12. Draw the phase trajectory of a free-falling particle of mass m.

13. What is the average position $< x >$ and the uncertainty Δx for a particle whose condition can be described by the following wavefunction:

$$\Psi(x) = N \exp\left(-\frac{1}{2}\lambda x^2\right).$$

   First, find the normalization factor N.

14. What are the wavelengths of a photon and an electron, each having kinetic energy $K = 1\,eV$?

15. What is the wavelength of a bullet of mass $m = 5\,g$ with a velocity $v = 1500\,ms^{-1}$?

16. The general expression for the average momentum of a particle moving in a single dimension is

$$< p >= \int_{-\infty}^{+\infty} \Psi(x)^* p \Psi(x)dx = -i\hbar \int_{-\infty}^{+\infty} \Psi(x)^* \Psi'(x)dx,$$

where $\Psi(x)^*$ is the complex conjugate of $\Psi(x)$, and $\Psi'(x) = d\Psi(x)/dx$.

a) Prove that if the wavefunction is real then $< p >= 0$.

b) Prove that $< p^2 >= \hbar^2 \int_{-\infty}^{+\infty} |\Psi'(x)|^2 dx$.

c) Calculate the uncertainty product $(\Delta x)(\Delta p)$ for the eigenfunction of the previous problem.

17. Plot the potential energy, wavefunctions, and probability densities for a particle in an infinite square well potential.

18. Determine the average position of a particle in a one-dimensional infinite square well.

19. One can show, solving Schrödinger's equation, that the energy of a moving particle of mass m, confined to a rectangular parallelepiped of lengths a, b, and c is, in the absence of external potential fields,

$$E_{n_x n_y n_z} = \frac{h^2}{8m} \left(\frac{n_x^2}{a^2} + \frac{n_y^2}{b^2} + \frac{n_z^2}{c^2} \right),$$

with $n_x, n_y, n_z = 1, 2, \ldots$.

Calculate the value of n_x, n_y, n_z for the case of a hydrogen atom (atomic weight $= 1.00$) in a box of dimensions $a = 1\,\text{cm}$, $b = 2\,\text{cm}$, $c = 3\,\text{cm}$ if the particle has kinetic energy $3k_B T/2$, at $T = 50°C$. What significant fact does this calculation illustrate?

20. Compute the ground-state energy for an electron that is confined to a potential well with a width of $L = 0.25\,\text{nm}$.

4

Ensemble theory

Josiah Willard Gibbs (1839–1903) developed the theory that provides an insightful, mechanistic understanding of thermodynamics. The theory accomplishes this by introducing the idea of a macroscopic thermodynamic state as an ensemble of microscopic states. Gibbs connected this ensemble of microscopic states to macroscopic properties by answering the following question: what is the observable difference between two different microscopic states, $\underline{X}_1$ and $\underline{X}_2$, that represent systems that are macroscopically identical? The answer is at the same time simple and profound: there is no observable difference.

Indeed in classical mechanics with a continuum phase space, there exists an infinitely large collection of microscopic systems that correspond to a particular macroscopic state. Gibbs named this collection of points in phase space an ensemble of systems. Gibbs then shifted the attention from trajectories, i.e., a succession of microscopic states in time, to all possible available state points in phase space that conform to given macroscopic, thermodynamic constraints. He then defined the probability of each member of an ensemble and determined thermodynamic properties as averages over the entire ensemble. In this chapter, we present the important elements of Gibbs' ensemble theory, setting the foundation for the rest of the book.

4.1 Distribution function and probability density in phase space

Consider an arbitrarily large collection $\mathcal{N}$ of systems with N particles ($\mathcal{N}$ and N are not to be confused), each pertaining to a separate point $\underline{X} = (\underline{p}, \underline{q})$ in $6N$-dimensional phase space, Γ. This manifold of points is called an ensemble.

Consider the points $\Delta \mathcal{N}$ in a given region of phase space (Fig. 4.1). Assuming that all points are equally probable, the probability of finding a point within this region is

$$\Delta P = \frac{\Delta \mathcal{N}}{\mathcal{N}}. \tag{4.1}$$

In the limit of an infinitesimal phase space region $d\underline{X} = d\underline{p}\,d\underline{q}$, from $\underline{X}$ to $\underline{X} + d\underline{X}$, the probability of finding a point in $d\underline{X}$ is

$$dP = \frac{d\mathcal{N}}{\mathcal{N}}. \tag{4.2}$$

We can choose $\mathcal{N}$ to be sufficiently large and populate the phase space densely enough to consider $d\mathcal{N}$ a continuously varying quantity.

We can then define the phase space probability distribution function as

$$D(\underline{X}) = \frac{d\mathcal{N}}{d\underline{X}}. \tag{4.3}$$

If we integrate the distribution function over the entire phase space we end up with the size of the ensemble

$$\mathcal{N} = \int_{\Gamma} D(\underline{X})\,d\underline{X}. \tag{4.4}$$

We can write for the differential probability of finding a microscopic state in $d\underline{X}$ that

$$dP = \frac{D(\underline{X})\,d\underline{X}}{\displaystyle\int_{\Gamma} D(\underline{X})\,d\underline{X}}. \tag{4.5}$$

Figure 4.1 Number of points $\Delta\mathcal{N}$ in a finite area of phase space.

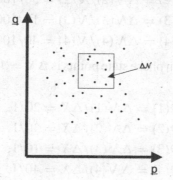

We define the phase space probability density $\rho(\underline{X})$, writing

$$dP = \rho(\underline{X})\,d\underline{X}, \tag{4.6}$$

where

$$\rho(\underline{X}) = \rho(\underline{p}, \underline{q}) = \frac{D(\underline{X})}{\displaystyle\int_{\Gamma} D(\underline{X})\,d\underline{X}}. \tag{4.7}$$

The probability density is defined so that it is independent of the number $\mathcal{N}$ assumed to constitute the ensemble.

The normalization condition is written as

$$\int_{\Gamma} \rho(\underline{X})d\underline{X} = 1. \tag{4.8}$$

Example 4.1
To illustrate these concepts, consider a very simple system that has a one-dimensional phase space of size 4. This means that the system can attain only one of four microscopic states. Let us assume that the possible microscopic states are $X = [1, 2, 3, 4]$. Assume an ensemble of $\mathcal{N} = 100$ systems that are distributed in phase space as follows:

$$\Delta \mathcal{N}(1) = 20,$$
$$\Delta \mathcal{N}(2) = 30,$$
$$\Delta \mathcal{N}(3) = 10,$$
$$\Delta \mathcal{N}(4) = 40. \tag{4.9}$$

Consequently, the probabilities of finding a system at any one of the four available points are

$$\Delta P(1) = \Delta \mathcal{N}(1)/\mathcal{N}(1) = 20/100,$$
$$\Delta P(2) = \Delta \mathcal{N}(2)/\mathcal{N}(2) = 30/100,$$
$$\Delta P(3) = \Delta \mathcal{N}(3)/\mathcal{N}(3) = 10/100,$$
$$\Delta P(4) = \Delta \mathcal{N}(4)/\mathcal{N}(4) = 40/100. \tag{4.10}$$

The interval between phase state points is $\Delta X = 1$ and the distribution function is discrete,

$$D(1) = \Delta \mathcal{N}(1)/\Delta X = 20/1,$$
$$D(2) = \Delta \mathcal{N}(2)/\Delta X = 30/1,$$
$$D(3) = \Delta \mathcal{N}(3)/\Delta X = 10/1,$$
$$D(4) = \Delta \mathcal{N}(4)/\Delta X = 40/1. \tag{4.11}$$

The probability density is

$$\rho(X) = \frac{\Delta P(X)}{\Delta X} = \begin{cases} 0.2, & X = 1 \\ 0.3, & X = 2 \\ 0.1, & X = 3 \\ 0.4, & X = 4 \end{cases}. \tag{4.12}$$

The probability distribution function and the probability density $\rho(X)$ should be independent of the ensemble size $\mathcal{N}$ in order to be properly defined.

4.2 Ensemble average of thermodynamic properties

Ensemble theory is the crowning achievement of Josiah Willard Gibbs. With ensemble theory he was able to provide for the first time a lucid connection between microscopic states and macroscopic observables. In this section we describe how. Consider any observable, thermodynamic or mechanical property M of a system with N particles. We have argued that any such property is a function of positions and momenta of the particles in the system,

$$M = M(\underline{X}) = M(\underline{p}, \underline{q}). \tag{4.13}$$

We define the ensemble average of this property as the average over a large collection $\mathcal{N}$ of macroscopically identical systems:

$$\langle M \rangle_{\text{ensemble}} = \frac{1}{\mathcal{N}} \sum_{n=1}^{\mathcal{N}} M(\underline{X}_n). \tag{4.14}$$

For arbitrarily large $\mathcal{N}$ we can define a continuous distribution function $D(\underline{X})$ and a continuous probability density $\rho(\underline{X})$.

The ensemble average of M can then be defined as

$$\langle M \rangle_{\text{ensemble}} = \frac{\int_{\Gamma} M(\underline{X}) D(\underline{X}) d\underline{X}}{\int_{\Gamma} D(\underline{X}) d\underline{X}}, \tag{4.15}$$

or, simplifying

$$\langle M \rangle_{\text{ensemble}} = \int_{\Gamma} M(\underline{X}) \rho(\underline{X}) d\underline{X}. \tag{4.16}$$

Example 4.2

Let us revisit the previous simple example and consider $M = M(X)$ to be a function of microscopic states. For example, let us assume that $M(X) = X^2$. The probability density is a discrete function and the ensemble average of M is given by

$$
\begin{aligned}
\langle M \rangle_{\text{ensemble}} &= \sum_{X=1}^{4} M(X)\rho(X) \\
&= 0.2M(1) + 0.3M(2) + 0.1M(3) + 0.4M(4) \\
&= 8.7.
\end{aligned}
\tag{4.17}
$$

4.3 Ergodic hypothesis

We have now presented two averages for macroscopic properties:

1. The trajectory average over consecutive time points:

$$
M_{\text{observed}} = \langle M \rangle = \lim_{T \to \infty} \frac{1}{T} \int_{0}^{T} M(\underline{X}(t))dt.
\tag{4.18}
$$

2. The ensemble average over distributed points in phase space:

$$
\langle M \rangle_{\text{ensemble}} = \int_{\Gamma} M(\underline{X})\rho(\underline{X})d\underline{X}.
\tag{4.19}
$$

A fundamental hypothesis in statistical thermodynamics is that the two averages are equal. This is the ergodic hypothesis. The physical implication of the ergodic hypothesis is that any system afforded with infinite time will visit all the points of phase space with a frequency proportional to their probability density. In other words, in a single trajectory the system spends an amount of time at each microstate that is proportional to its probability. The ergodic hypothesis is still a hypothesis because there has been no formal proof of its truth.

As observed in Chapter 3, trajectories never intersect themselves, except when the system periodically passes through each state. The practical difficulty is that for large systems, for example with Avogadro's number, N_A, particles, the time required for the system to visit the entire phase space and return to an initial microstate can be estimated to be very many times the age of the universe. The difficulties discussed when introducing the classical mechanical formalisms emerge again then when attempting to prove the ergodic hypothesis.

Nonetheless, the ergodic hypothesis has logical, physical underpinnings. If one does not accept the ergodic hypothesis, one is faced with logical inconsistencies in the physical description of matter. Importantly, the adoption of the ergodic hypothesis results in a correct prediction of thermodynamic quantities of matter. This is the ultimate test for any hypothesis or theory.

It is the ergodic hypothesis that allowed Gibbs to shift attention from trajectories to probabilities in phase space. Instead of considering orbits of microstate chains crossing the phase space in time, one can envision the phase space as a continuum with a position-dependent density. Because the latter can be determined more readily than the former, statistical thermodynamics can be employed to connect microscopic to macroscopic states.

4.4 Partition function

We can now introduce a new important property, called the partition function of the ensemble, which is generally defined as

$$Z = \frac{1}{h^{3N} N!} \int_{\Gamma} D(\underline{X}) d\underline{X}. \tag{4.20}$$

The partition function gives a measure of the number of members in the ensemble that give rise to any specific macroscopic thermodynamic state. Besides the probability distribution and the probability in phase space, equilibrium ensembles can also be described with the help of the partition function.

We explain the prefactor term $\frac{1}{h^{3N} N!}$ in Eq. 4.20 in the following section.

4.5 Microcanonical ensemble

Consider an ensemble of systems, each with constant number of particles N, in constant volume V, with constant energy E. This is called the NVE or microcanonical ensemble. Each member system of this ensemble corresponds to a point $\underline{X} = (\underline{p}, \underline{q})$ in the $6N$-dimensional phase space Γ.

The question we will answer is the following: how are these systems distributed in phase space? More precisely, we will determine the phase space probability density, $\rho_{NVE}(\underline{X})$.

Since all the systems have an energy E, the points $\underline{X}$ are bound on a constant energy phase space hypersurface, defined as

$$\Sigma(N, V, E) = \int_{p_1} \int_{p_2} \cdots \int_{p_{3N}} \int_{q_1} \int_{q_2} \cdots \int_{q_{3N}}$$
$$\times \delta(H(p_1, p_2, \ldots, p_{3N}, q_1, q_2, \ldots, q_{3N}) - E)$$
$$\times dp_1\, dp_2 \ldots dp_{3N}\, dq_1\, dq_2, \ldots dq_{3N}, \qquad (4.21)$$

or, more compactly

$$\Sigma(N, V, E) = \int_\Gamma \delta(H(\underline{X}) - E)\, d\underline{X}. \qquad (4.22)$$

In Equation 4.22, Σ is the size of phase space occupied by the NVE ensemble of microscopic states. The Dirac δ function simply selects the NVE points and includes them in Σ. In other words, all points that represent systems with energy other than E are not included in the integral. The area Σ is then a $(6N - 1)$-dimensional surface, because not all degrees of freedom can change independently while satisfying the constant energy constraint.

Equation 4.22 implies that the phase space is a continuum, where microscopic states can be defined infinitely close to one another. As discussed in Chapter 3, this physical picture is unrealistic. Quantum mechanics dictates a discrete phase space with each microscopic state occupying an area equal to h^{3N}, one uncertainty measure for each position–momentum pair.

From the size of the available phase space $\Sigma(N, V, E)$ we can define the number of microscopic states $\Omega(N, V, E)$ that comprise the NVE ensemble. Simply,

$$\Omega(N, V, E) = \frac{\Sigma(N, V, E)}{h^{3N} N!}, \qquad (4.23)$$

or

$$\Omega(N, V, E) = \frac{1}{h^{3N} N!} \int_\Gamma \delta(H(\underline{X}) - E)\, d\underline{X}. \qquad (4.24)$$

The available phase space volume, $\Sigma(N, V, E)$, when divided by h^{3N}, becomes a discrete, non-dimensional number, Ω, which quite simply turns out to be the size of the ensemble. Because of quantum mechanics, the number of microscopic states corresponding to a particular NVE thermodynamic state is finite, albeit incomprehensibly large, as we discuss later in this chapter when we determine Ω for an ideal gas.

Note that $\Sigma(N, V, E)$ is also divided by $N!$, which is the number of microstates that are identical for N indistinguishable particles, as discussed in Chapter 2. This term $N!$, accounting for identical ways that indistinguishable particles can be placed in space, is called the Boltzmann correction. It is interesting to note that Ludwig Boltzmann added the $N!$ term in order to obtain the correct thermodynamic properties, without much physical intuition. A physical picture was actually unavailable in the late 1890s, when Boltzmann added his correction, because the concepts of a discrete phase space with indistinguishable particles were yet undeveloped. It was later, in the 1930s, with the work of John Kirkwood that the physical meaning of correction terms became clear.

We can now determine the probability density at each point in a continuum phase space as follows:

$$\rho_{NVE}(\underline{X}) = \frac{\delta(H(\underline{X}) - E)}{\Sigma(N, V, E)} = \begin{cases} \dfrac{1}{\Sigma(N, V, E)}, & \text{if } H(\underline{X}) = E \\ 0 & \text{otherwise} \end{cases} \quad (4.25)$$

This simply means that the NVE probability density is uniform everywhere in the available area of the phase space, or that all microscopic state points of an NVE ensemble are of equal probability.

In a discrete phase space the probability of each state is determined as follows:

$$P_{NVE}(\underline{X}) = \frac{\delta(H(\underline{X}) - E)}{\Omega(N, V, E)} = \begin{cases} \dfrac{1}{\Omega(N, V, E)}, & \text{if } H(\underline{X}) = E \\ 0 & \text{otherwise} \end{cases} \quad (4.26)$$

This means that all possible phase space points $\underline{X}$ that correspond to a particular NVE macroscopic state have the same probability of occurring.

The size of the available phase space, Σ, and the number of microscopic states, Ω, corresponding to the particular NVE state can be determined in principle for any system with known Hamiltonian. We determine them for an ideal gas later in this chapter.

4.6 Thermodynamics from ensembles

Generally, with properties determined, such as Ω and the probability density, there are two methods we can use to derive thermodynamic properties:

1. The first method makes use of the definition of ensemble averages for any macroscopic, mechanical property $M(\underline{X})$ of an NVE system. The quantity $M(\underline{X})$ can be determined as an NVE ensemble average as follows:

$$\langle M(\underline{X}) \rangle_{NVE} = \int_{\Gamma} M(\underline{X}) \rho_{NVE}(\underline{X}) \, d\underline{X}. \qquad (4.27)$$

2. The second method uses Boltzmann's equation for the entropy as a starting point. Boltzmann was able to determine the entropy of the NVE ensemble with the following equation, first presented in Chapter 1:

$$S(N, V, E) = k_B \ln \Omega(N, V, E). \qquad (4.28)$$

With Eq. 4.28 at hand, all other thermodynamic properties can be determined, such as the temperature, the pressure, or the chemical potential. We can start with the fundamental relation for the energy of any system, which is derived by combining the first and second laws of thermodynamics:

$$dE = T dS - P dV + \mu dN. \qquad (4.29)$$

Using Boltzmann's equation, Eq. 4.28, we can write

$$\left. \frac{\partial S}{\partial E} \right|_{N,V} = \frac{1}{T} \qquad (4.30)$$

or

$$\left. \frac{\partial \ln(\Omega)}{\partial E} \right|_{N,V} = \frac{1}{k_B T}. \qquad (4.31)$$

Similarly, by differentiating the entropy with respect to the volume

$$\left. \frac{\partial S}{\partial V} \right|_{N,E} = \left. \frac{\partial \ln(\Omega)}{\partial V} \right|_{N,E}$$

$$= \frac{P}{k_B T}. \qquad (4.32)$$

Finally, by differentiating the entropy with respect to the number of particles, we can determine the particle chemical potential of an NVE system:

$$\left. \frac{\partial S}{\partial N} \right|_{E,V} = \left. \frac{\partial \ln(\Omega)}{\partial N} \right|_{E,V}$$

$$= -\frac{\mu}{k_B T}. \qquad (4.33)$$

4.7 $S = k_B \ln \Omega$, or entropy understood

In Chapter 1, we presented a method proposed by Einstein to prove the relation between the entropy S and the number of microscopic states Ω. At this point, it may be instructive to present another way to reach Boltzmann's equation.

Consider two separate equilibrium physical NVE systems A_1 with (N_1, V_1, E_1) and A_2 with (N_2, V_2, E_2) (Fig. 4.2). The number of corresponding microscopic states is $\Omega_1(N_1, V_1, E_1)$ and $\Omega_2(N_2, V_2, E_2)$ respectively.

Consider that the two systems are brought into thermal contact allowing for energy exchange, but allowing for no mass exchange or change of the volume of the systems. Then, the two energies E_1 and E_2 become variable under the restriction that

$$E_0 = E_1 + E_2 = \text{constant}, \tag{4.34}$$

where E_0 is the energy of the total, composite system A_0.

Figure 4.2 Separate NVE systems in thermal contact (top panel), with movable boundaries (middle), and with porous boundaries (bottom).

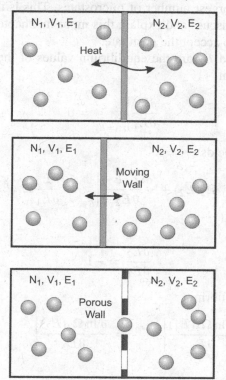

The composite system is an NVE system, equally likely to be in any of Ω_0 microstates. Because system A_1 can attain any microscopic state independent of the microscopic state attained by system A_2, we can write

$$\Omega_0 = \Omega_1(E_1)\Omega_2(E_2)$$
$$= \Omega_1(E_1)\Omega_2(E_0 - E_1)$$
$$= \Omega_0(E_0, E_1). \tag{4.35}$$

This means that during the process of energy exchange between the two systems A_1 and A_2, the number of microscopic states of the composite system Ω_0 changes with E_1. An important question then arises: at what value of variable E_1 will the composite system be at equilibrium, or, in other words, how far will the energy exchange proceed?

Let us postulate that the final, equilibrium energy, E_1, is the one that maximizes Ω_0. The logic behind this assertion is that a system left alone proceeds naturally in the direction that enables it to assume an ever increasing number of microstates, until it settles down to the macrostate that affords the largest number of microstates. This is the equilibrium macrostate. We discuss and explain this more in Chapter 6, but at this point it suffices to accept the postulate.

Assume $\bar{E}_1$ and $\bar{E}_2$ are the equilibrium values of the energy in the two systems. Then

$$\left.\frac{\partial \Omega_0}{\partial E_1}\right|_{E_1=\bar{E}_1} = 0. \tag{4.36}$$

Using Eq. 4.35 yields

$$\left.\frac{\partial \Omega_1(E_1)}{\partial E_1}\right|_{E_1=\bar{E}_1} \Omega_2(\bar{E}_2) + \left.\frac{\partial \Omega_2(E_2)}{\partial E_2}\right|_{E_2=\bar{E}_2} \frac{\partial E_2}{\partial E_1}\Omega_1(\bar{E}_1) = 0. \tag{4.37}$$

Since

$$\frac{\partial E_2}{\partial E_1} = -1 \tag{4.38}$$

we obtain at equilibrium

$$\left.\frac{\partial \ln \Omega_1(E_1)}{\partial E_1}\right|_{E_1=\bar{E}_1} = \left.\frac{\partial \ln \Omega_2(E_2)}{\partial E_2}\right|_{E_2=\bar{E}_2}. \tag{4.39}$$

If we define a new variable

$$\beta = \frac{\partial \ln \Omega(E)}{\partial E}\bigg|_{E=\bar{E}}, \tag{4.40}$$

the equilibrium condition reduces to

$$\beta_1 = \beta_2. \tag{4.41}$$

It is known empirically that the energy exchange will continue between two systems in thermal contact until their temperatures become equal, $T_1 = T_2$. Parameter β is indeed directly related to the temperature. If we use the thermodynamic relationship

$$\frac{\partial S}{\partial E}\bigg|_{N,V} = \frac{1}{T} \tag{4.42}$$

at equilibrium for constant N and V for both systems we obtain

$$\frac{\partial S_1}{\partial E_1}\bigg|_{E_1=\bar{E}_1} = \frac{\partial S_2}{\partial E_2}\bigg|_{E_2=\bar{E}_2}. \tag{4.43}$$

Comparing Eq. 4.40 and Eq. 4.42 we can write

$$\beta = \frac{1}{k_B T}, \tag{4.44}$$

where k_B is a constant.

We also obtain, once again, Boltzmann's equation

$$S = k_B \ln \Omega. \tag{4.45}$$

Different, more rigorous proofs are available in the books found in Further reading.

The logical implication of this discussion is that systems exchange energy so that the entropy of the composite system becomes maximum at equilibrium. The entropy is directly connected to the number of microscopic states, attaining an intuitive statistical character. Naturally, a macrostate with the largest possible number of microstates is the most probable one.

Besides thermal equilibrium, we can expand this discussion to mechanical and chemical equilibria, as follows. Assume that the separation wall between the two systems A_1 and A_2 is a movable piston. We know empirically that the piston will move until the systems reach mechanical equilibrium. Although the volumes of each of the two systems can change, the total volume of the composite system

$$V_0 = V_1 + V_2 \tag{4.46}$$

is constant.

Let us denote with $\bar{V}_1$ and $\bar{V}_2$ the equilibrium volumes of systems A_1 and A_2, respectively. Asserting again that V_1 changes until the number of microscopic states of the composite system Ω_0 becomes maximum at equilibrium we can write, following the same approach as before,

$$\left. \frac{\partial \Omega_0}{\partial V_1} \right|_{V_1 = \bar{V}_1} = 0. \tag{4.47}$$

Then in both systems

$$\left. \frac{\partial \Omega_1(V_1)}{\partial V_1} \right|_{V_1 = \bar{V}_1} \Omega_2(\bar{V}_2) + \left. \frac{\partial \Omega_2(V_2)}{\partial V_2} \right|_{V_2 = \bar{V}_2} \frac{\partial V_2}{\partial V_1} \Omega_1(\bar{V}_1) = 0. \tag{4.48}$$

Since

$$\frac{\partial V_2}{\partial V_1} = -1, \tag{4.49}$$

we obtain at equilibrium for constant N and E

$$\left. \frac{\partial \ln \Omega_1(V_1)}{\partial V_1} \right|_{V_1 = \bar{V}_1} = \left. \frac{\partial \ln \Omega_2(V_2)}{\partial V_2} \right|_{V_2 = \bar{V}_2}. \tag{4.50}$$

Combining Eq. 4.38, Eq. 4.50 and the thermodynamic relationship

$$\left. \frac{\partial S}{\partial V} \right|_{N,E} = \frac{P}{T} \tag{4.51}$$

eventually yields that the two systems reach mechanical equilibrium when the pressures are equal, that is $P_1 = P_2$.

Finally, if A_1 and A_2 are brought into contact allowing particles to move and exchange between the two systems, following the same steps yields for equilibrium that the chemical potential is equal in both systems, that is $\mu_1 = \mu_2$.

Therefore, starting from the assertion that systems move to equilibrium maximizing the number of corresponding microstates and using Boltzmann's equation, we recover the criteria

$$
\begin{aligned}
T_1 &= T_2, \\
P_1 &= P_2, \\
\mu_1 &= \mu_2,
\end{aligned}
\tag{4.52}
$$

for thermal, mechanical, and chemical equilibria.

It is a fundamental dictum of classical thermodynamics that an isolated system, with constant mass, volume, and energy, attains equilibrium when its entropy reaches its maximum value. It is also well established that equilibrium can be operationally defined as the macroscopic state with no internal system gradients in temperature, pressure, and chemical potential.

With ensemble theory, we consider all the possible microscopic states that correspond to the NVE macroscopic state, and we can define the entropy of the system in concretely intuitive terms. Importantly, instead of asserting a maximum entropy value at equilibrium, we realize the equilibrium is the macrostate that simply affords the largest number of microscopic states.

4.8 Ω for ideal gases

We can now derive the thermodynamic properties of an ideal gas starting from microscopic properties. In particular the procedure of connecting microscopic and thermodynamic properties typically has the following steps:

1. Determine the Hamiltonian of the system. Knowledge of the molecular level details is needed at this stage, i.e., how atoms interact and how molecules are formed.
2. Determine the ensemble (e.g., NVE or NVT) to work with, depending on constraints. Remember that for single-component, single-phase systems two constraints and the number of particles uniquely define the equilibrium state and all thermodynamic properties.
3. Determine the probability density, ρ, or the partition function as a function of the Hamiltonian and the ensemble constraints. This is a taxing task, requiring elements of probability and combinatorial theory.
4. Determine thermodynamic properties. In principle, this can be accomplished with two approaches:
   a) determine properties using ensemble averaging with the help of the probability density, or the partition function;
   b) alternatively, use a relation between a thermodynamic property and the partition function. For example, we have seen that $S = k_B \ln \Omega$ for an NVE system. Similar relations will be developed for other ensembles in the next two chapters.

Let us follow these steps for an NVE ideal gas.

1. In an ideal gas, the particles do not interact and the potential energy of the system is $U(\underline{q}) = 0$. The Hamiltonian of an NVE ideal gas of

particles with mass m is then simply the kinetic energy

$$H(\underline{p}, \underline{q}) = \sum_i^N \frac{p_i^2}{2m},\tag{4.53}$$

where $p_i^2 = p_{ix}^2 + p_{iy}^2 + p_{iz}^2$ is the squared momentum of particle i.

2. The probability density in the phase space of an NVE ideal gas (IG) is then

$$\rho_{NVE-IG}(\underline{X}) = \frac{\delta\left(\sum_i^N \frac{p_i^2}{2m} - E\right)}{\int_\Gamma \delta\left(\sum_i^N \frac{p_i^2}{2m} - E\right) d\underline{X}}.\tag{4.54}$$

3. The partition function is

$$\Omega_{IG}(N, V, E) = \frac{1}{h^{3N} N!} \int_\Gamma \delta\left(\sum_i^N \frac{p_i^2}{2m} - E\right) d\underline{p} d\underline{q}.\tag{4.55}$$

Since the Hamiltonian does not depend on the positions of the particles we can integrate the δ function in the integrand over the positions separately,

$$\Omega_{IG}(N, V, E) = \frac{1}{h^{3N} N!} \int_\Gamma d\underline{q} \int_\Gamma \delta\left(\sum_i^N \frac{p_i^2}{2m} - E\right) d\underline{p}.\tag{4.56}$$

For a system constrained in volume V, considering Cartesian coordinates and assuming a cubic box with sides of size L, so that $L^3 = V$, the first integral in Eq. 4.56 is

$$\int_\Gamma d\underline{q} = \int_{q_1} \int_{q_2} \cdots \int_{q_N} d\underline{q}_1 d\underline{q}_2 \cdots d\underline{q}_N$$

$$= \int_0^L dx_1 \int_0^L dy_1 \int_0^L dz_1 \int_0^L dx_2 \int_0^L dy_2$$

$$\times \int_0^L dz_2 \cdots \int_0^L dx_N \int_0^L dy_N \int_0^L dz_N$$

$$= V^N.\tag{4.57}$$

This result is proportional to the number of configurational microscopic states, that is, the number of arrangements of N particles in volume V. The division by $N!$ in Eq. 4.55 ensures the proper counting of ways to arrange indistinguishable particles.

The second integral in Eq. 4.56 results in the size of a $(3N - 1)$-dimensional hypersurface. Let us present the calculation inductively.

First, consider a simple constant energy system with only two degrees of freedom, p_1 and p_2 (this could be a single particle confined on a flat surface) (Fig. 4.3). The Hamiltonian is

$$H = \frac{p_1^2}{2m} + \frac{p_2^2}{2m} = \frac{p^2}{2m} = E. \tag{4.58}$$

The available phase space is equal to the perimeter, $2\pi p$, of a circle with radius $p = \sqrt{p_1^2 + p_2^2} = \sqrt{2mE}$.

Consider now a simple constant energy system with only three degrees of freedom, p_1, p_2, and p_3 (this could be a single particle confined in a box) (Fig. 4.4). The Hamiltonian is

$$H = \frac{p_1^2}{2m} + \frac{p_2^2}{2m} + \frac{p_3^2}{2m} = \frac{p^2}{2m} = E. \tag{4.59}$$

Figure 4.3 The available phase space of a constant energy classical system with two degrees of freedom, p_1 and p_2, is the perimeter of a circle of radius $p = \sqrt{p_1^2 + p_2^2} = \sqrt{2mE}$.

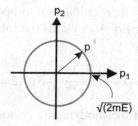

Figure 4.4 The available phase space of a constant energy classical system with three degrees of freedom, p_1, p_2, and p_3, is the surface of a sphere of radius $p = \sqrt{p_1^2 + p_2^2 + p_3^3} = \sqrt{2mE}$.

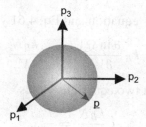

The available phase space is equal to the surface, $4\pi p^2$ of a sphere of radius $p = \sqrt{p_1^2 + p_2^2 + p_3^3} = \sqrt{2mE}$.

Considering an ideal gas with N particles, there are $3N$ momenta. Writing the Hamiltonian again

$$H = \frac{p_1^2}{2m} + \frac{p_2^2}{2m} + \cdots \frac{p_{3N}^2}{2m} = \frac{p^2}{2m} = E. \qquad (4.60)$$

The available phase space is equal to the size S_{3N} of the hyper-surface determined as the integral over $(3N - 1)$ angles, multiplied by $p^{3N-1}(2mE)^{3N/2}$. Note that we are making the approximation $3N - 1 \approx 3N$. This is reasonable for systems with large numbers of particles.

Note that the constant S_{3N} is only a function of the dimensionality of the phase space, which in turn is determined by the number of particles.

We can now compute the total number of microscopic states Ω for an NVE ideal gas. Multiplying the results for the configurational and the momenta parts of the phase space yields the following relation for Ω:

$$\Omega_{IG}(N, V, E) = \frac{V^N}{h^{3N} N!} S_{3N} (2mE)^{3N/2}. \qquad (4.61)$$

4. The determination of thermodynamic properties is at this point straightforward. Using Eq. 4.45 yields the following relation for the entropy of the ideal gas

$$S_{IG}(N, V, E) = k_B \ln \left[\frac{V^N}{h^{3N} N!} S_{3N} (2mE)^{3N/2} \right]. \qquad (4.62)$$

From the fundamental thermodynamic relation for the energy of a system we get

$$\left. \frac{\partial S}{\partial V} \right|_{E,N} = \frac{P}{T}. \qquad (4.63)$$

Using Boltzmann's equation and Eq. 4.61 yields

$$k_B \left. \frac{\partial \ln \Omega}{\partial V} \right|_{E,N} = \frac{k_B N}{V}. \qquad (4.64)$$

Combining the last two equations yields

$$\frac{k_B N}{V} = \frac{P}{T}. \qquad (4.65)$$

Similarly, when we differentiate the entropy with respect to the energy we obtain the following:

$$\frac{\partial S}{\partial E}\Big|_{V,N} = \frac{1}{T}. \tag{4.66}$$

Using Boltzmann's equation

$$k_B \frac{\partial \ln \Omega}{\partial E}\Big|_{V,N} = \frac{3k_B N}{2E}. \tag{4.67}$$

Combining the last two equations yields

$$\frac{3k_B N}{2E} = \frac{1}{T}. \tag{4.68}$$

Rearranging Eqs. 4.65 and 4.68 we obtain

$$PV = Nk_BT \tag{4.69}$$

and

$$E = \frac{3}{2}Nk_BT. \tag{4.70}$$

Comparing with the empirical ideal gas law,

$$PV = nRT = \frac{N}{N_A}RT, \tag{4.71}$$

the Boltzmann constant is $k_B = R/N_A = 1.38 \times 10^{-23} \text{JK}^{-1}$.

We can now take some stock. We have just derived the celebrated ideal gas equation of state and determined the energy of a gas as a function of the temperature. We accomplished this by starting with the microscopic Hamiltonian and by employing ensemble theory concepts. Thus, we provided a concrete connection between microscopic and macroscopic properties for ideal gases. In the following few chapters we continue deriving the thermodynamics of different systems starting with microscopic arguments.

It is worth noting that a purely classical mechanical representation of ideal gas particles yields accurate results. In other words, the thermodynamic behavior of ideal gases is fully describable without any mention of quantum mechanics. We discuss how in the next section.

4.9 Ω with quantum uncertainty

The correct result for Ω_{NVE-IG} is in actuality slightly different from the one in Eq. 4.61. In order to be precise we have to take into account quantum mechanical uncertainty. Remarkably, the final derived thermo-dynamic properties remain the same.

Quantum mechanical uncertainty dictates that the system actually has energy with values between E and $E + \delta E$, where the uncertainty in the energy is much smaller than the energy itself, $\delta E \ll E$.

The calculation of the configurational part of the phase space remains the same, but instead of calculating the size of a $(3N - 1)$-dimensional hypersurface for the momenta, we need to compute the size of a $3N$-dimensional hypersurface with the final dimension being of thickness δE (Fig. 4.5). In other words we need to determine the number of cells for which the Hamiltonian $H(\underline{X})$, or more simply the kinetic energy $K(\underline{p})$, falls in a narrow range of values $E \leq K(\underline{p}) \leq E + \delta E$.

More concisely, we write

$$\Omega_{NVE-IG} = \frac{V^N}{h^{3N} N!} \int_{E < K < E + \delta E} d\underline{p}. \tag{4.72}$$

Following a similar process, we first compute the size of phase space for $N = 1$ confined in three dimensions (Fig. 4.6). This is the volume of a spherical shell with

$$\sqrt{2mE} \leq p \leq \sqrt{2m(E + \delta E)}. \tag{4.73}$$

Figure 4.5 The available phase space of a constant energy system with two degrees of freedom, p_1 and p_2, is the area between two circles of radius $\sqrt{2mE}$ and $\sqrt{2m(E + \delta E)}$.

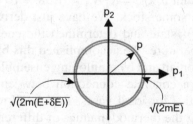

Figure 4.6 The available phase space of a constant energy system with three degrees of freedom, p_1, p_2, and p_3, is the volume between two spheres of radius $\sqrt{2mE}$ and $\sqrt{2m(E + \delta E)}$.

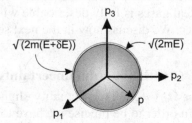

The volume of the shell for $\delta E \ll E$ is equal to

$$4\pi(2mE)\delta p, \tag{4.74}$$

where

$$\delta p = \sqrt{2m(E + \delta E)} - \sqrt{2mE} \tag{4.75}$$

and

$$\delta p \approx \frac{d}{dE}(\sqrt{2mE})\delta E \tag{4.76}$$

or, finally

$$\delta p \approx \frac{\delta E}{2E}\sqrt{2mE}. \tag{4.77}$$

The shell volume is then

$$2\pi\frac{\delta E}{E}(2mE)^{3/2}. \tag{4.78}$$

Another way to derive this is by calculating the integral

$$\int_{E < K(\underline{p}) < E + \delta E} d\underline{p} = 4\pi \int_{\sqrt{2mE}}^{\sqrt{2m(E+\delta E)}} p^2 dp. \tag{4.79}$$

After some algebraic manipulation we obtain the shell volume

$$\int_{E < K(\underline{p}) < E + \delta E} d\underline{p} = 2\pi\frac{\delta E}{E}(2mE)^{3/2}. \tag{4.80}$$

For N particles, the momentum is

$$p = \left(\sum_{i=1}^{N} |p_i|^2\right)^{1/2}. \tag{4.81}$$

Instead of 4π, which is the integral over two angles for a 3-dimensional sphere, we now have the S_{3N} integral over $3N - 1$ angles. Then

$$\int_{E < K(\underline{p}) < E + \delta E} d\underline{p} = S_{3N} \int_{\sqrt{2mE}}^{\sqrt{2m(E+\delta E)}} p^{3N-1} dp. \tag{4.82}$$

Similar to the three-dimensions, algebraic manipulation yields

$$\int_{E < K(\underline{p}) < E + \delta E} d\underline{p} = S_{3N}\frac{\delta E}{E}(2mE)^{3N/2}. \tag{4.83}$$

Finally, the number of microscopic states that correspond to the same thermodynamic NVE state is

$$\Omega_{IG}(N, V, E) = \frac{V^N}{h^{3N} N!} S_{3N} (2mE)^{3N/2} \frac{\delta E}{E}. \tag{4.84}$$

It can be quickly verified that although the results in Eq. 4.61 and Eq. 4.82 are different by a factor $\delta E / E$, the derived thermodynamic properties of an ideal gas remain the same. This means that thermodynamic properties do not depend on the exact number of microscopic states, or more generally the actual value of the partition function. Instead, they depend on the functional dependence of the partition function on NVE properties and how it changes with the ensemble constraints.

Note also that the actual value of Ω is not computable for a large number of degrees of freedom. Even the most powerful computer cannot produce a number for Ω for a system with Avogadro's number of particles. Indeed, the value for Ω is incomprehensibly large. This is one reason why a precise, absolute numerical value for the entropy at a certain NVE state may not be meaningful.

A second reason is that Eq. 4.62 assumes that at absolute zero temperature the energy and the entropy are zero, $E = 0$ and $S = 0$. This is not true because of quantum uncertainty at $T = 0\,\mathrm{K}$, which results in $\Omega(T = 0\,\mathrm{K}) \neq 1$.

There is, in other words, a residual entropy, S_0 at absolute zero that must be added to the entropy of Eq. 4.62. We decide instead to work with relative entropies, choosing a ground state where we define $S = 0$.

A third reason is that in actuality the subatomic degrees of freedom result in an expansion of the phase space and an increase of microscopic states. Nevertheless, for the types of system we are interested in here, these contributions are considered constant, only altering the reference state of thermodynamic properties.

4.10 Liouville's equation

Liouville derived a remarkable equation that succinctly captures the dynamics of physical systems, describing the evolution of the probability density in time, both at and away from equilibrium.

Consider an ensemble $\mathcal{N}$ of points in phase space, each representing a system of N particles. As time evolves each microscopic state moves along a phase trajectory according to the equations of motion. The entire ensemble of points then resembles the particles of a fluid in motion.

Let $\mathcal{V}$ be an arbitrary $6N$-dimensional closed domain in Γ (Fig. 4.7). Consider $\mathcal{S}$ as the $(6N - 1)$-dimensional surface of this domain. Define $\underline{v}(p, q)$ to be the vector locally normal to $\mathcal{S}$ at $\underline{X} = (p, q)$. Thinking of space points as fluid particles, the phase space density is now a function of both position and time, $\rho = \rho(\underline{X}, t)$.

The continuity principle holds in this case, as there is no creation or destruction of systems. The implication is that, when considering a closed phase space domain $\mathcal{V}$, the net number of points leaving $\mathcal{S}$ per unit time equals the rate of decrease of points in volume $\mathcal{V}$. This continuity principle can be expressed as follows:

$$\int_{\mathcal{S}} \rho \left(\underline{v} \cdot \underline{\dot{X}} \right) dS = -\frac{\partial}{\partial t} \int_{\mathcal{V}} \rho \, d\mathcal{V}, \tag{4.85}$$

where $\underline{\dot{X}}$ is the velocity of microscopic state-points moving in phase space.

According to the divergence theorem, the left-hand side in Eq. 4.85 can be written as follows:

$$\int_{\mathcal{S}} \rho \left(\underline{v} \cdot \underline{\dot{X}} \right) dS = \int_{\mathcal{V}} \text{div} \left(\rho \underline{\dot{X}} \right) d\mathcal{V}, \tag{4.86}$$

where the divergence in phase space is defined as

$$\text{div} \left(\rho \underline{\dot{X}} \right) = \sum_{i=1}^{3N} \left[\frac{\partial}{\partial q_i} (\rho \dot{q}_i) + \frac{\partial}{\partial p_i} (\rho \dot{p}_i) \right]. \tag{4.87}$$

Figure 4.7 Trajectories flow through the area, $\mathcal{S}$, of an arbitrary $6N$-dimensional closed domain, $\mathcal{V}$.

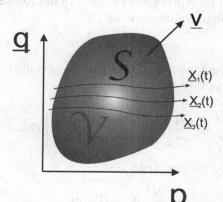

Equation 4.85 now becomes

$$\int_{\mathcal{V}} \left[\text{div}(\rho \underline{X}) + \frac{\partial \rho}{\partial t} \right] d\mathcal{V} = 0. \tag{4.88}$$

For Eq. 4.88 to be valid for any arbitrary phase space domain $\mathcal{V}$, the following must be true:

$$\text{div}(\rho \, \underline{X}) + \frac{\partial \rho}{\partial t} = 0. \tag{4.89}$$

This is the continuity equation in phase space, analogous to the continuity equation of fluids.

Using Hamilton's equations of motion, it is trivial to show that

$$\text{div}(\rho \, \underline{X}) = \sum_{i=1}^{3N} \left[\frac{\partial H}{\partial p_i} \frac{\partial}{\partial q_i} - \frac{\partial H}{\partial q_i} \frac{\partial}{\partial p_i} \right] \rho. \tag{4.90}$$

Finally then, we obtain

$$\frac{\partial \rho}{\partial t} + \sum_{i=1}^{3N} \left[\frac{\partial H}{\partial p_i} \frac{\partial}{\partial q_i} - \frac{\partial H}{\partial q_i} \frac{\partial}{\partial p_i} \right] \rho = 0. \tag{4.91}$$

This is Liouville's equation. It completely specifies the evolution of the probability density for a system of a specified Hamiltonian. It is also equivalent to $6N$ Hamilton's equations of motion.

Systems at equilibrium have a constant probability density, that is

$$\frac{\partial \rho}{\partial t} = 0. \tag{4.92}$$

Thus, the equilibrium probability density of microstates only depends on the Hamiltonian at equilibrium, i.e.,

$$\rho = \rho(H). \tag{4.93}$$

We can also write Liouville's equation for the probability distribution $D(\underline{X}) = D(\underline{p}, \underline{q})$ as follows:

$$\frac{\partial D}{\partial t} + \sum_{i=1}^{3N} \left[\frac{\partial H}{\partial p_i} \frac{\partial}{\partial q_i} - \frac{\partial H}{\partial q_i} \frac{\partial}{\partial p_i} \right] D = 0. \tag{4.94}$$

More compactly, we can write

$$\frac{\partial D}{\partial t} + i\mathcal{L}D = 0, \tag{4.95}$$

where $i\mathcal{L}$ is the Liouville operator, defined as

$$i\mathcal{L} = \sum_{i=1}^{3N} \left[\frac{\partial H}{\partial p_i} \frac{\partial}{\partial q_i} - \frac{\partial H}{\partial q_i} \frac{\partial}{\partial p_i} \right]. \tag{4.96}$$

Another way to write the Liouville equation is by using the Poisson bracket operator $\{H, \dots\}$

$$\frac{\partial D}{\partial t} + \{H, D\} = 0. \tag{4.97}$$

The Liouville and the Poisson bracket operators are related as follows

$$i\mathcal{L} = \{H, \dots\}. \tag{4.98}$$

Liouville's equation is the starting point for our discussion of non-equilibrium systems in Chapter 12.

4.11 Further reading

1. J. W. Gibbs, *Elementary Principles in Statistical Mechanics*, (New Haven: Yale University Press, 1902).
2. D. A. McQuarrie, *Statistical Mechanics*, (Sausalito, CA: University Science Books, 2000).
3. T. L. Hill, *Statistical Thermodynamics*, (Reading, MA: Addison-Wesley, 1960).
4. F. Reif, *Statistical and Thermal Physics*, (New York: McGraw-Hill, 1965).
5. R. C. Tolman, *Statistical Mechanics*, (London: Oxford University Press, 1938).
6. J. E. Mayer and M. G. Mayer, *Statistical Mechanics*, (New York: Wiley, 1940).
7. D. ter Haar, *Elements of Statistical Mechanics*, (New York: Renehart, 1954).
8. F. C. Andrews, *Equilibrium Statistical Mechanics*, (New York: John Wiley and Sons, 1963).

4.12 Exercises

1. Demonstrate that Hamilton's equations of motion are invariant under a reversal of the direction of time $t \to (-t)$. What are the implications for a non-equilibrium system approaching equilibrium in positive times under constant NVE conditions? Is this invariance contrary to the second law of thermodynamics?

2. Find the entropy $S(NVE)$ of an ideal gas of N classical particles, with an energy E contained in a d-dimensional box of volume V. Derive the equation of state, assuming that N is very large.

3. Determine the number of microstates $\Omega(N, V, E)$ for an ideal gas with number density $\rho = N\sigma^3/V = 0.6$, where $\sigma = 1$ is the size of the particles, which

have mass $m = 10^{-24}$ g. Consider a temperature $T = 250\,$K, for i) $N = 10$, ii) $N = 1000$, iii) $N = 10^6$, and iv) $N = 10^9$. Comment on the size of the available space in each case. Can $\Omega(N, V, E)$ actually be calculated for 1 mole of particles?

4. Consider a system of N particles belonging to a fictitious ensemble we call the "uniform ensemble." The probability density of the uniform ensemble is indeed uniform for all states with energy equal to or less than E. Find the partition function of the uniform ensemble.

5. Using the microcanonical ensemble, compute the Helmholtz free energy $A(T, N)$ as a function of temperature for a system of N identical but distinguishable particles, each of which has two possible energy levels, $\pm\varepsilon$.

6. Consider a classical harmonic oscillator. What is the size, Σ, of the phase space that is occupied by the oscillator when its energy, E, is constant? Considering quantum mechanical uncertainty and the consequent discretization of phase space, what is the number of microscopic states Ω? What are the entropy, enthalpy, pressure, temperature, and Gibbs free energy of the oscillator?

7. What are Σ and Ω, if the energy of the classical harmonic oscillator can take values between E_1 and E_2 $(E_1 \leq E \leq E_2)$?

8. For a collection of N three-dimensional classical harmonic oscillators of frequency ω and fixed total energy E, compute the entropy S and the temperature T. Should the oscillators be treated as distinguishable or indistinguishable? What are the entropy, enthalpy, pressure, temperature, and Gibbs free energy of the oscillator collection?

9. Consider an ideal gas system with N particles in volume V at a constant temperature T. Assume that the microscopic state of this system changes every 10^{-30} seconds, i.e., the particles move enough during that interval for the point in phase space to be considered different. Convince yourself that the probability of finding a microscopic state where all the particles are confined in a tenth of the total volume is the same as any other microscopic state. Then determine the average time it will take for the system to be observed in such a state: in other words, how many years on average should you wait to observe all the particles in a specific corner with a volume equal to $1/10$ of the total volume?

5

Canonical ensemble

Consider a system of N particles in volume V. Assume that the temperature T of the system is constant. The ensemble of microscopic points in phase space that correspond to this NVT macroscopic state is called the NVT or canonical ensemble. The canonical ensemble is called that simply because it was the first ensemble tackled by Gibbs. Indeed, ensemble theory as described in Gibbs' seminal text (see Further reading) was formulated entirely for the constant temperature ensemble. The reason, as will become apparent in this chapter, is the ease in determining the canonical partition function.

In this chapter, we present an illustrative calculation of the probability density in phase space and the partition function of the canonical ensemble. We then again derive ideal gas thermodynamics.

5.1 Probability density in phase space

In order for the temperature of a system S to remain constant, the system must be in thermal contact with a heat bath B (Fig. 5.1). Assume that the system has energy E_S and the heat bath has energy E_B. The composite system is isolated with a constant total energy $E = E_B + E_S$. The bath can be arbitrarily large. Consequently, we may assume $E_B \gg E_S$.

Since the system and the heat bath are in thermal contact, their temperatures will be the same at equilibrium,

$$T_S = T_B = T. \tag{5.1}$$

Energy can flow freely between the system and the heat bath. Consequently E_S is no longer fixed, but can fluctuate.

Assume that system S can attain any one of a large number of different energy levels $E_i, i = 1, 2, \ldots,$ with probability $p_S(E_i)$. For each energy level E_i, the system can be in any of $\Omega(E_i)$ microscopic states. We will use a counter $j = 1, 2, \ldots, \Omega(E_i)$ for each of the microstates at each energy level E_i.

Each microscopic state that the system can attain will have a probability p_{ij}. Obviously, the probabilities of microstates with the same energy are equal:

$$p_{i1} = p_{i2} = \cdots = p_{i\Omega(E_i)}. \tag{5.2}$$

We can write

$$p_{ij} = \frac{1}{\Omega(E_i)} p_S(E_i), \quad j = 1, 2, \ldots, \Omega(E_i) \tag{5.3}$$

where, again, $p_S(E_i)$ is the probability of system S having an energy E_i.

We need to determine what $p_S(E_i)$ is. We can prove, following any one of the very many proofs available in the literature, that

$$p_S(E_i) \propto \exp(-\beta E_i), \tag{5.4}$$

where

$$\beta = \frac{1}{k_B T}. \tag{5.5}$$

Here, we will prove Eq. 5.4 in the following illustrative way. In the following chapter we will discuss a different proof. Let us consider two independent systems, A and B, with N_A, V_A, T and N_B, V_B, T, respectively, inside a heat bath. At equilibrium the temperature will be the same in both systems.

The probability that system A will be at a particular energy level E_A is $P_A(E_A)$. Similarly the probability that system B will be at a particular energy level E_B is $P_B(E_B)$.

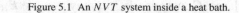

Figure 5.1 An NVT system inside a heat bath.

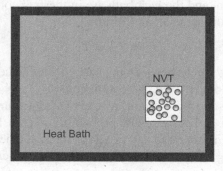

Consider now the composite system, $A + B$. The probability that the composite system is at an energy level $E_A + E_B$ is

$$P_{A+B}(E_A + E_B) = P_A(E_A) P_B(E_B). \tag{5.6}$$

Differentiating Eq. 5.6 with respect to $E_A + E_B$ yields

$$\frac{dP_{A+B}(E_A + E_B)}{d(E_A + E_B)} = \frac{d(P_A(E_A) P_B(E_B))}{dE_A} \frac{dE_A}{d(E_A + E_B)}$$
$$= \frac{d(P_A(E_A) P_B(E_B))}{dE_B} \frac{dE_B}{d(E_A + E_B)}. \tag{5.7}$$

Since Eq. 5.7 is valid for arbitrarily chosen systems, we deduce that for any system in a heat bath

$$\frac{1}{P_A(E_A)} \frac{dP_A(E_A)}{dE_A} = \frac{1}{P_B(E_B)} \frac{dP_B(E_B)}{dE_B} = -\beta, \tag{5.8}$$

where β must be a constant. We will prove in the next section that indeed $\beta = 1/k_B T$, but let us turn back to calculating $p_S(E_i)$.

Equation 5.8 is true for system S, which simply means that

$$p_S(E_i) \propto \exp(-\beta E_i). \tag{5.9}$$

Combining Eq. 5.3 with Eq. 5.9 results trivially in

$$p_{ij} \propto \exp(-\beta E_i). \tag{5.10}$$

Use of the probability normalization condition results in the following expression for the probability p_{ij}:

$$p_{ij} = \frac{\exp(-\beta E_i)}{\sum_j \sum_i \exp(-\beta E_i)}, \tag{5.11}$$

where the sums run over all i energy levels E_i and over all j microscopic states with energy E_i.

Consequently, the probability of microstates belonging to the canonical, NVT, ensemble is

$$p_{ij} = \frac{\exp(-\beta E_i)}{\sum_i \Omega(N, V, E_i) \exp(-\beta E_i)}. \tag{5.12}$$

Finally, we can define the partition function of the canonical ensemble as

$$Q(N, V, T) = \sum_i \Omega(N, V, E_i) \exp(-\beta E_i)$$

$$= \sum_k \exp(-\beta E_k), \qquad (5.13)$$

where the sum over k runs over the entire ensemble of microscopic states in phase space that are available to the NVT system.

Using a continuum, classical mechanical approach, the probability density of each microscopic point $\underline{X} = (\underline{p}, \underline{q})$ is determined as

$$\rho(\underline{X}) = \frac{\exp(-\beta H(\underline{X}))}{\displaystyle\int_\Gamma \exp(-\beta H(\underline{X})) d\underline{X}}. \qquad (5.14)$$

The canonical partition function can now be written as

$$Q(N, V, T) = \frac{1}{h^{3N} N!} \int_\Gamma \exp(-\beta H(\underline{X})) d\underline{X}, \qquad (5.15)$$

where the term $1/(h^{3N} N!)$ is added to take into account the discrete nature of phase space and the indistinguishability of identical particles, all described in the previous chapter.

The canonical ensemble average of any property M is then

$$\langle M \rangle_{NVT} = \frac{\displaystyle\int_\Gamma M(\underline{X}) \exp(-\beta H(\underline{X})) d\underline{X}}{\displaystyle\int_\Gamma \exp(-\beta H(\underline{X})) d\underline{X}}. \qquad (5.16)$$

Reverting to a discrete state space, the ensemble average of M is written as

$$\langle M \rangle_{NVT} = \frac{\displaystyle\sum_k M(k) \exp(-\beta E(k)))}{\displaystyle\sum_k \exp(-\beta E(k))}, \qquad (5.17)$$

where $M(k)$ is the value of the property in state k.

5.2 *NVT* ensemble thermodynamics

With the partition function at hand, we can derive expressions for all important thermodynamic properties.

The energy of a system with constant NVT fluctuates. The ensemble average is the macroscopically observable internal energy, U, with

$$
\begin{aligned}
U &= \langle E \rangle_{NVT} \\
&= \sum_k E_k p_k \\
&= \sum_k \frac{E_k \, e^{-\beta E_k}}{Q},
\end{aligned}
\tag{5.18}
$$

where, again, the sum over k runs over the entire ensemble of microscopic states in phase space that are available to the NVT system.

In Eq. 5.18, the partition function Q is determined as follows:

$$
Q = \sum_k e^{-\beta E_k}.
\tag{5.19}
$$

Therefore, we can write

$$
U = -\frac{\partial \ln Q}{\partial \beta}\bigg|_{N,V}.
\tag{5.20}
$$

Classical thermodynamics instructs us that

$$
U = A + TS,
\tag{5.21}
$$

where A is the Helmholtz free energy of the system.

The fundamental relationship for A is the following (see Appendix B):

$$
dA = -S dT - P dV + \mu dN.
\tag{5.22}
$$

Therefore the entropy can be written as

$$
S = -\frac{\partial A}{\partial T}\bigg|_{N,V}.
\tag{5.23}
$$

Then Eq. 5.21 becomes

$$
U = A - T \frac{\partial A}{\partial T}\bigg|_{N,V}
\tag{5.24}
$$

or, more compactly

$$U = \left.\frac{\partial(A/T)}{\partial(1/T)}\right|_{N,V}. \tag{5.25}$$

Comparing Eq. 5.25 to Eq. 5.20 we can conclude that

$$\beta = \frac{1}{k_B T} \tag{5.26}$$

and, importantly, that

$$A = -k_B T \ln Q. \tag{5.27}$$

The introduction of the constant k_B in Eq. 5.27 and consequently in Eq. 5.26 is needed to properly account for property units. This constant is Boltzmann's constant.

The canonical variables of the Helmholtz free energy for closed systems are the temperature T and volume V, as is apparent in Eq. 5.22. A canonical ensemble system is at equilibrium then when the number of available microscopic states is maximum, that is when the partition function is maximum, or equivalently when its Helmholtz free energy is minimum.

With the canonical partition function $Q(N, V, T)$ we can also calculate other thermodynamic properties, such as the pressure.

We can start by rewriting the internal energy U in terms of the temperature

$$U = k_B T^2 \left.\frac{\partial \ln Q}{\partial T}\right|_{N,V}. \tag{5.28}$$

Equation 5.28 is a powerful expression, yielding the internal energy of any system of constant NVT. Later in this chapter we use it to determine the internal energy of an ideal gas. Then in subsequent chapters, we use it to derive the internal energy of non-ideal gases, liquids, and solids, among others.

The fundamental thermodynamics expression for the internal energy is

$$dU = T dS - P dV + \mu dN, \tag{5.29}$$

leading to the following result for the pressure:

$$P = k_B T \left.\frac{\partial \ln Q}{\partial V}\right|_{N,T}. \tag{5.30}$$

5.3 Entropy of an *NVT* system

For a fixed number of particles N, we can start with the fundamental thermodynamic relation for the internal energy and solve for the infinitesimal change in entropy. This yields

$$dS = \frac{1}{T}(dU + P dV). \tag{5.31}$$

From Eq. 5.18

$$dU = \sum_k (p_k dE_k + E_k dp_k). \tag{5.32}$$

The significance of the foregoing expression is that the energy of a system may change either as a result of a shift in the available energy levels or as a result of a change in the probability of each energy level. The first term corresponds to work done on the system, whereas the second term is associated with thermal changes in the system, because of heat exchange with a heat bath (there is an extensive discussion on this in advanced statistical mechanics texts, but it is beyond the scope of the present text).

Indeed, when the system is at a microscopic state k, we can write for the work done on the system by a volume change

$$dE_k = -P_k dV, \tag{5.33}$$

where P_k is the pressure of microscopic state k.

Equation 5.32 now becomes

$$dU = \sum_k [p_k P_k dV + E_k dp_k]. \tag{5.34}$$

Using the definition for the ensemble average of the pressure gives

$$dU = -P dV + \sum_k E_k dp_k. \tag{5.35}$$

For the entropy, Eq. 5.31 becomes

$$dS = \frac{\left(-P dV + \sum_k E_k \, dp_k + P dV \right)}{T}, \tag{5.36}$$

and

$$dS = \frac{\sum_k E_k \, dp_k}{T}. \tag{5.37}$$

For the canonical ensemble, by definition the microstate probability is

$$p_k = \frac{e^{-\beta E_k}}{\sum\limits_k e^{-\beta E_k}}. \tag{5.38}$$

Solving for the energy yields

$$E_k = -\frac{1}{\beta} \ln p_k - \frac{1}{\beta} \ln Q. \tag{5.39}$$

Substituting for the energy in Eq. 5.37 yields

$$dS = -\frac{\sum\limits_k (\ln p_k \, dp_k + \ln Q dp_k)}{\beta T}. \tag{5.40}$$

Of course,

$$\sum_k p_k = 1, \tag{5.41}$$

which means that

$$\sum_k dp_k = 0. \tag{5.42}$$

Consequently, and since $\ln Q$ is a constant,

$$dS = -k_B \sum_k \ln p_k dp_k. \tag{5.43}$$

Assuming that at a temperature of absolute zero $S(T = 0\,\text{K}) = 0$, we can integrate the entropy from $T = 0\,\text{K}$ to any temperature $T > 0$. This simply yields

$$S = \frac{U}{T} + k_B \ln Q \tag{5.44}$$

or, alternatively

$$S = k_B T \left. \frac{\partial \ln Q}{\partial T} \right|_{N,V} + k_B \ln Q. \tag{5.45}$$

This, of course, is a result we could have obtained using Eq. 5.23. Consequently, it is straightforward to show that

$$S = -k_B \sum_k p_k \ln p_k \tag{5.46}$$

and, finally

$$S = -k_B \langle \ln(p_k) \rangle_{NVT}. \tag{5.47}$$

According to Eq. 5.47, the entropy is calculated as a weighted average of microscopic state probabilities. In other words, entropy is nothing but a measure of the spread of probabilities in phase space.

It is instructive to summarize what we have derived so far for the microcanonical ensemble and the canonical ensemble.

Partition function

NVE ensemble: $\Omega(N, V, E) = \dfrac{1}{h^{3N} N!} \displaystyle\int_\Gamma \delta(H(\underline{X}) - E)\, d\underline{X}$

NVT ensemble: $Q(N, V, T) = \dfrac{1}{h^{3N} N!} \displaystyle\int_\Gamma \exp(-\beta H(\underline{X}))\, d\underline{X}$

Ensemble average

NVE ensemble: $\langle M(\underline{X}) \rangle_{NVE} = \dfrac{\displaystyle\int_\Gamma M(\underline{X}) \delta(H(\underline{X}) - E)\, d\underline{X}}{\displaystyle\int_\Gamma \delta(H(\underline{X}) - E)\, d\underline{X}}$

NVT ensemble: $\langle M(\underline{X}) \rangle_{NVT} = \dfrac{\displaystyle\int_\Gamma M(\underline{X}) \exp(-\beta H(\underline{X}))\, d\underline{X}}{\displaystyle\int_\Gamma \exp(-\beta H(\underline{X}))\, d\underline{X}}$

Canonical thermodynamic property

NVE ensemble: $S(N, V, E) = k_B \ln \Omega(N, V, E)$

NVT ensemble: $A(N, V, T) = -k_B T \ln Q(N, V, T)$

5.4 Thermodynamics of *NVT* ideal gases

We can determine the thermodynamic properties of an ideal gas using canonical ensemble statistical mechanical calculations. We follow the same steps as in the NVE ideal gas. The Hamiltonian of an NVT ideal gas of particles with mass m is only a function of the momenta

$$H(\underline{p}, \underline{q}) = \sum_{i=1}^{3N} \frac{p_i^2}{2m}, \tag{5.48}$$

where p_i is one of the three momenta for each of the N particles.

The partition function is

$$Q(N, V, T) = \frac{1}{h^{3N} N!} \int d\underline{q} \int d\underline{p} \exp(-\beta H(\underline{p}, \underline{q})). \tag{5.49}$$

Again, as in Chapter 4, the $3N$ integrals $\int d\underline{q}$ over the configurational part of the phase space yield the volume of the system for each particle. Furthermore, using Eq. 5.48 we can write

$$Q(N, V, T) = \frac{V^N}{h^{3N} N!} \prod_{i=1}^{3N} \int dp_i \exp\left(-\beta \frac{p_i^2}{2m}\right). \tag{5.50}$$

The integral for each of the momenta degrees of freedom can be conveniently calculated with the help of the following formula:

$$\int_{-\infty}^{+\infty} \exp(-\phi^2) d\phi = \sqrt{\pi}. \tag{5.51}$$

Any quadratic term in the Hamiltonian can then be easily computed in the NVT probability density calculation with Eq. 5.51.

For the ideal gas, we make the substitution

$$\phi = \sqrt{\frac{\beta}{2m}} p_i \tag{5.52}$$

to yield for the partition function

$$Q(N, V, T) = \frac{V^N}{N!} \left(\frac{2\pi m k_B T}{h^2}\right)^{3N/2}. \tag{5.53}$$

We can define a new property called the thermal wavelength, which is constant for a specific type of particle at constant temperature, as follows:

$$\Lambda = \left(\frac{h^2}{2\pi m k_B T}\right)^{1/2}. \tag{5.54}$$

The partition function for an NVT ideal gas is concisely written then as

$$Q(N, V, T) = \frac{V^N}{\Lambda^{3N} N!}. \tag{5.55}$$

Although an attempt to calculate the actual value of the partition function for a system with $N = 1$ mole of particles is absurd, this simple and elegant formula can be used to determine theoretically the thermodynamic behavior of an ideal gas.

Using Stirling's approximation, the absolute Helmholtz free energy is

$$A = -k_B T \ln Q$$
$$= -Nk_B T \left(1 + \ln \left(\frac{V}{N} \left(\frac{2\pi m k_B T}{h^2} \right)^{3/2} \right) \right). \tag{5.56}$$

The absolute entropy can be computed as

$$S = - \left. \frac{\partial A}{\partial T} \right|_{N,V}$$
$$= Nk_B \left(\frac{5}{2} + \ln \left(\frac{V}{N} \left(\frac{2\pi m k_B T}{h^2} \right)^{3/2} \right) \right). \tag{5.57}$$

The internal energy is then

$$U = A + TS$$
$$= \frac{3}{2} Nk_B T. \tag{5.58}$$

The constant volume heat capacity can now be determined as

$$C_V = - \left. \frac{\partial U}{\partial T} \right|_{N,V}$$
$$= \frac{3}{2} Nk_B$$
$$= \frac{3}{2} nR, \tag{5.59}$$

where $n = N/N_A$ is the number of moles and R is the ideal gas constant.

The molar heat capacity $\underline{C}_V$ is simply

$$\underline{C}_V = \frac{3}{2} \frac{nR}{n}$$
$$= \frac{3}{2} R. \tag{5.60}$$

The pressure of an ideal gas can be determined using the Helmholtz free energy fundamental relation and differentiating with respect to the volume as follows:

$$P = -\frac{\partial A}{\partial V}\bigg|_{T,N}. \tag{5.61}$$

Using Eq. 5.56 yields

$$P = \frac{Nk_BT}{V}. \tag{5.62}$$

Rearranging gives

$$PV = Nk_BT, \tag{5.63}$$

or in terms of numbers of moles

$$PV = nRT. \tag{5.64}$$

The chemical potential of an ideal gas is

$$\mu = \frac{\partial A}{\partial N}\bigg|_{V,T}$$

$$= -k_BT \ln \left(\frac{V}{N} \left(\frac{2\pi m k_B T}{h^2} \right)^{3/2} \right). \tag{5.65}$$

In macroscopic thermodynamics, the chemical potential of an ideal gas is typically written as

$$\mu(P, T) = \mu^o(T) + k_BT \ln P, \tag{5.66}$$

where $\mu^o(T)$ is the standard chemical potential, which depends only on the temperature.

Combining Eq. 5.65 and Eq. 5.66 yields for the standard chemical potential

$$\mu^o = -k_BT \ln \left(k_BT \left(\frac{2\pi m k_B T}{h^2} \right)^{3/2} \right). \tag{5.67}$$

The ideal gas law and the thermodynamic properties of an ideal gas are completely derived from the canonical ensemble partition function. This is a remarkable illustration of how statistical mechanics explains macroscopic observables in terms of microscopic properties.

We are now in a position to use these calculations to derive the thermodynamic behavior of systems other than the ideal gas. We will explore non-ideal gases, liquids, and simple models of solids, for which the partition function can be determined analytically.

The theoretical derivation provides physical insight into the nature of thermodynamic properties such as the free energy and the chemical potential. We discuss these further in the following chapters.

We will also explain why two different ensembles, the microcanonical, NVE, and the canonical, NVT, yield the same thermodynamic behavior for ideal gases, and indeed for any macroscopic system.

5.5 Calculation of absolute partition functions is impossible and unnecessary

It is noteworthy that a statistical mechanical calculation of absolute values of entropy or free energy is not required for determination of thermodynamic properties of matter. The functional dependence of the partition function on macroscopic properties, such as the total mass, volume, and temperature of the system, is sufficient to derive equations of state, internal energies, and heat capacities. For example, knowledge that the ideal gas partition function scales as V^N is adequate to define and explain the ideal gas equation of state.

Entropies, free energies, and chemical potentials are typically needed in relative terms, as differences between distinct macroscopic states. For example, the sign of the entropy change provides information on the direction in which a change of state process will spontaneously move. Nonetheless, it is impossible to determine an actual value for the absolute entropy or free energy of a macroscopic system. There are at least two reasons.

First, from a physical perspective, it is not feasible to account for absolutely all degrees of freedom of a system in calculation of a partition function. Taking into account all subatomic degrees of freedom is impractical. Approximations such as the Born–Oppenheimer approximation that the nucleus of atoms is at the ground state are then indispensable, but the constant contribution to the total partition function is arbitrary.

Second, calculation of the actual value of a partition function is intractable. For an ideal gas, the NVT ensemble partition function is

$$Q(N, V, T) = \frac{V^N}{\Lambda^{3N} N!}. \tag{5.68}$$

Even for as small a number of particles as 100, the actual value of Q can be computed and found to be an incomprehensibly large number.

Again, all that is needed to derive macroscopic thermodynamics from microscopic relations is the functional dependence of the partition function on ensemble constraints.

5.6 Maxwell–Boltzmann velocity distribution

The Hamiltonian of an ideal gas with N particles can be written as

$$H(\underline{p}_1, \underline{p}_2, \cdots, \underline{p}_N, \underline{q}_1, \underline{q}_2, \cdots, \underline{q}_N) = \sum_{i=1}^{N} h(\underline{p}_i, \underline{q}_i), \quad (5.69)$$

where

$$h(\underline{p}_i, \underline{q}_i) = \frac{\underline{p}_i^2}{2m} \quad (5.70)$$

is the one-particle Hamiltonian.

The partition function can be written as

$$Q(N, V, T) = \frac{1}{N! h^{3N}} \int d\underline{p} d\underline{q} \exp\left(-\beta \sum_{i=1}^{3N} h(\underline{p}_i, \underline{q}_i)\right) \quad (5.71)$$

or

$$Q(N, V, T) = \frac{1}{N! h^{3N}} \prod_{i=1}^{N} \int d\underline{p}_i d\underline{q}_i \exp(-\beta h(\underline{p}_i, \underline{q}_i)). \quad (5.72)$$

The partition function can be interpreted as

$$Q(N, V, T) = \frac{q(1, V, T)^N}{N!}, \quad (5.73)$$

where

$$q(1, V, T) = \frac{1}{h^3} \prod_{i=1}^{N} \int d\underline{p}_i d\underline{q}_i \exp(-\beta h(\underline{p}_i, \underline{q}_i)) \quad (5.74)$$

is the one-particle partition.

The single-particle partition function for an ideal gas is easily found to be

$$q(1, V, T) = \frac{V}{\Lambda^3}. \quad (5.75)$$

The phase space density for the NVT system is

$$\rho_N = \frac{\exp(-\beta H(\underline{p}, \underline{q}))}{Q(N, V, T)},$$ (5.76)

or, alternatively

$$\rho_N = N! \prod_{i=1}^{N} \frac{\exp(-\beta h(\underline{p}_i, \underline{q}_i))}{q_i(1, N, T)}.$$ (5.77)

The probability density of finding a single particle at $(\underline{p}_i, \underline{q}_i)$ is

$$\rho_i(\underline{p}_i, \underline{q}_i) = \frac{\exp(-\beta h(p_i, q_i))}{q_i(1, N, T)},$$ (5.78)

or

$$\rho_i(\underline{p}_i, \underline{q}_i) = \frac{\Lambda^3}{V} \exp(-\beta h(p_i, q_i)).$$ (5.79)

Consequently, the probability of finding a particle within $(\underline{q}_i, \underline{p}_i)$ and $(\underline{q}_i + d\underline{q}_i, \underline{p}_i + d\underline{p}_i)$ is

$$\rho_i d\underline{p}_i d\underline{q}_i = \frac{\Lambda^3}{V} \exp(-\beta h(p_i, q_i)) d\underline{p}_i d\underline{q}_i.$$ (5.80)

We can go from the probability distribution in phase space, $\rho_i d\underline{p}_i d\underline{q}_i$ to the velocity distribution of particles in volume V, $f(\underline{u})d\underline{u}$ simply by changing the variables from momenta to velocities and integrating out the positions. Finally,

$$f(\underline{u}) = \left(\frac{m}{2k_B T} \right)^{3/2} \exp \left(-\frac{mu^2}{2k_B T} \right).$$ (5.81)

This is the celebrated Maxwell–Boltzmann distribution.

Since it is difficult to illustrate $f(\underline{u})$ graphically, in Fig. 5.2 we plot the distribution of the x component of the velocity for various gases, calculated with the expression

$$f(u_x) = \left(\frac{m}{2k_B T} \right)^{3/2} \exp \left(-\frac{mu_x^2}{2k_B T} \right).$$ (5.82)

The average velocity is zero, with negative and positive velocities being equally probable. If this were not true, then the total system momentum in the x direction would not be zero, and the center of mass of the entire system would be moving.

The probability distribution of speeds, that is of velocity magnitudes, $u = |\underline{u}|$ can be derived from $f(\underline{u})$ integrating over angles. The integration yields

$$f(u) = 4\pi u^2 \left(\frac{m}{2\pi k_B T} \right)^{3/2} \exp \left(-\frac{mu^2}{2k_B T} \right). \tag{5.83}$$

In Fig. 5.3, the Maxwell–Boltzmann distribution for oxygen is shown at various temperatures.

Figure 5.2 Maxwell–Boltzmann one-component velocity distribution for various gases at 300 K.

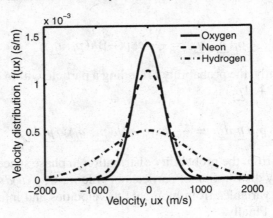

Figure 5.3 Maxwell–Boltzmann speed distribution for oxygen at various temperatures.

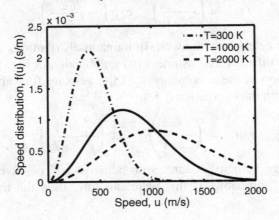

5.7 Further reading

1. J. W. Gibbs, *Elementary Principles in Statistical Mechanics*, (New Haven: Yale University Press, 1902).
2. D. A. McQuarrie, *Statistical Mechanics*, (Sausalito, CA: University Science Books, 2000).
3. T. L. Hill, *Statistical Thermodynamics*, (Reading, MA: Addison-Wesley, 1960).
4. F. Reif, *Statistical and Thermal Physics*, (New York: McGraw-Hill, 1965).
5. R. C. Tolman, *Statistical Mechanics*, (London: Oxford University Press, 1938).
6. J. E. Mayer and M. G. Mayer, *Statistical Mechanics*, (New York: Wiley, 1940).
7. D. ter Haar, *Elements of Statistical Mechanics*, (New York: Renehart, 1954).
8. F. C. Andrews, *Equilibrium Statistical Mechanics*, (New York: John Wiley and Sons, 1963).
9. D. Chandler, *Introduction to Modern Statistical Mechanics*, (New York: Oxford University Press, 1987).
10. T. L. Hill, *Statistical Mechanics: Principles and Selected Applications*, (New York: Dover, 1987).
11. J. D. Walecka, *Fundamentals of Statistical Mechanics*, (Stanford, CA: Stanford University Press, 1989).

5.8 Exercises

1. Use Eq. 5.55 to calculate the partition function of argon. Assume the following conditions:
   a) $N = 10, T = 300$ K, $V = 10$ m^3;
   b) $N = 10^3, T = 300$ K, $V = 10^{-20}$ m^3;
   c) $N = 10^6, T = 300$ K, $V = 10^{-20}$ m^3;
   d) $N = 10^6, T = 300$ K, $V = 10$ m^3;
   e) $N = 10^{12}, T = 300$ K, $V = 10$ m^3.

2. For the systems considered in the previous problem, calculate the Helmholtz free energy, the entropy, the internal energy, the heat capacity, and the pressure. Compare with literature experimental values of these properties. A useful resource is the National Institute of Standards and Technology (NIST) Web-Book (available at *webbook.nist.gov/chemistry*).

3. Consult the National Institute of Standards and Technology (NIST) WebBook (available at *webbook.nist.gov/chemistry*) and find the heat capacity, enthalpy, and chemical potential for neon, xenon, hydrogen, and oxygen. Calculate these properties at the canonical ensemble and compare the results.

4. Using the canonical ensemble, compute the Helmholtz free energy $A(T, N)$ as a function of temperature for a system of N identical but distinguishable particles, each of which has two possible energy levels, $\pm\varepsilon$.

5. Consider a collection of N hard rods of size l, confined in one dimension in the interval $(0, L)$ with constant temperature T. Assume no translational motion.

Let the rods interact through a hard, pairwise potential

$$u(|x_i - x_j|) = \begin{cases} \infty, & \text{if } |x_i - x_j| < l \\ 0, & \text{if } |x_i - x_j| > l \end{cases}$$

where x_i and x_j are the positions of rods i and j, respectively. Prove that the partition function is

$$Q(N, L, T) = \frac{(L - (N - 1)l)^N}{N!}.$$

6. Consider the NVT ensemble of an ideal gas that has a single-particle Hamiltonian of the form

$$h_i = a_i |p_i|^A + b_i |q_i|^B,$$

where a_i, b_i are constants and A, B are positive integers.
   What is the free energy of this system? What is its internal energy?

7. Consider the partition function Q_1 for an ensemble of systems of N_1 particles in volume V_1 at temperature T, and the partition function Q_2 for the ensemble of systems of N_2 particles in volume V_2 at the same temperature T. Find Q_2 if Q_1, V_1, and V_2 are known and if $N_2 = N_1(V_2/V_1)$.

8. A gas of N classical point particles lives on a two-dimensional surface of a sphere of radius r. Find the internal energy $U(T, A)$, where A is the surface area. Find the equation of state for this gas.

9. For an ideal binary mixture start with the canonical partition function and show that the internal energy is

$$U = \frac{3}{2}(N_1 + N_2)k_B T$$

and the entropy is

$$S = N_1 k_B \ln\left(\frac{V \exp(5/2)}{\Lambda_1^3 N_1}\right) + N_2 k_B \ln\left(\frac{V \exp(5/2)}{\Lambda_2^3 N_2}\right).$$

Then derive the standard thermodynamic formula for the entropy of mixing for an ideal gas mixture.

10. An ideal gas consisting of N particles of mass m (classical statistics is being obeyed) is enclosed in an infinitely tall cylindrical container of cross section s, placed in a uniform gravitational field, and is in thermal equilibrium. Assume that the potential energy $U = 0$ at $z = 0$. Calculate the classical partition function, Helmholtz free energy, mean energy, and heat capacity of this system.

11. An approximate partition function for a dense gas is of the form

$$Q(N, V, T) = \frac{1}{N!} \left(\frac{2\pi mkT}{h^2}\right)^{3N/2} (V - bN)^N e^{aN^2/VkT},$$

where *a* and *b* are constant molecular parameters. Calculate the equation of state from this partition function. What equation of state is it? Calculate the thermodynamic energy and the heat capacity and compare them to the ideal gas ones.

12. What fraction of oxygen molecules at 275.15 K have a speed within ±1% of the average speed?

Fluctuations and other ensembles

6.1 Fluctuations and equivalence of different ensembles

The equation of state for an ideal gas is the same when derived in the microcanonical ensemble and in the canonical ensemble. Indeed, all thermodynamic properties are determined to be the same regardless of the ensemble used. How can the thermodynamic behavior of a system with constant energy be the same as the behavior of a system whose energy fluctuates? To address this question we determine the magnitude of energy fluctuations in the canonical ensemble.

For an ensemble of equivalent NVT macroscopic systems, the ensemble averaged energy is

$$U = \langle E \rangle_{NVT}. \tag{6.1}$$

By definition, the variance of the energy is

$$\langle (\delta E)^2 \rangle = \langle (E - \langle E \rangle)^2 \rangle \tag{6.2}$$

or

$$\langle (\delta E)^2 \rangle = \langle E^2 - 2E \langle E \rangle + \langle E \rangle^2 \rangle, \tag{6.3}$$

which yields

$$\langle (\delta E)^2 \rangle = \langle E^2 \rangle - \langle E \rangle^2. \tag{6.4}$$

Note that this expression is valid for any fluctuating variable. We can write

$$\langle E^2 \rangle = \frac{\sum_j E_j^2 \exp(-\beta E_j)}{Q}, \tag{6.5}$$

where the sum runs over all available microscopic states j in the NVT ensemble.

Simple algebraic manipulations yield

$$\langle E^2 \rangle = -\frac{1}{Q} \frac{\partial}{\partial \beta} \sum_j E_j \exp(-\beta E_j). \qquad (6.6)$$

Consequently,

$$\langle E^2 \rangle = -\frac{1}{Q} \frac{\partial}{\partial \beta} (\langle E \rangle Q). \qquad (6.7)$$

More algebra yields

$$\langle E^2 \rangle = -\frac{\partial \langle E \rangle}{\partial \beta} - \langle E \rangle \frac{\partial \ln Q}{\partial \beta} \qquad (6.8)$$

and finally

$$\langle E^2 \rangle = k_B T^2 \left. \frac{\partial \langle E \rangle}{\partial T} \right|_{N,V} + \langle E \rangle^2. \qquad (6.9)$$

The variance of energy fluctuation is therefore

$$\langle (\delta E)^2 \rangle = k_B T^2 \left. \frac{\partial \langle E \rangle}{\partial T} \right|_{N,V}, \qquad (6.10)$$

or, more concisely

$$\langle (\delta E)^2 \rangle = k_B T^2 C_V. \qquad (6.11)$$

This is a remarkable result. In principle, a system at constant temperature will have a fluctuating internal energy. These fluctuations in energy are related to the rate at which energy changes with changes in temperature.

What is of interest is the magnitude of the energy fluctuations compared to the energy itself. We can estimate this relative magnitude of energy fluctuations for any NVT systems as follows:

$$\frac{\langle (\delta E)^2 \rangle^{1/2}}{\langle E \rangle} = \frac{(k_B T^2 C_V)^{1/2}}{\langle E \rangle}. \qquad (6.12)$$

For an ideal gas we know that

$$\langle E \rangle = \frac{3}{2} N k_B T \qquad (6.13)$$

and

$$C_V = \frac{3}{2} N k_B. \qquad (6.14)$$

Consequently, we can conclude that in a constant temperature ideal gas system the ratio of the standard deviation of the energy fluctuations over the average energy scales as

$$\frac{\langle\langle\delta E\rangle\rangle}{\langle E\rangle} \propto N^{-1/2}. \tag{6.15}$$

Equation 6.15 suggests that when the number of particles is large the fluctuations become insignificant. In particular, at the theoretical limit of $N \to \infty$, fluctuations vanish. This limit is called the thermodynamic limit.

Practically, for 1 mole of particles the number is large enough for the fluctuations to be considered negligible. Indeed it is difficult to observe energy differences when the energy standard deviation is approximately 0.000000001% of the average (Fig. 6.1). For a given temperature then, there is one measurable energy, the average energy of the NVT ensemble.

Note that this argument can be extended to systems beyond ideal gases that are at the thermodynamic limit. The proof is beyond the scope of this book.

Accordingly, at the thermodynamic limit, the probability of members of the NVT ensemble having a particular energy level E_i reduces to the following expression:

$$P(E_i) = \delta(E_i - \langle E\rangle). \tag{6.16}$$

Generally, the canonical partition function is

$$Q(N, V, T) = \sum_j \exp\left(\frac{-E_j}{k_B T}\right), \tag{6.17}$$

Figure 6.1 Fluctuations of the energy, E, of an NVT ensemble are negligible around the ensemble average, $\langle E\rangle$.

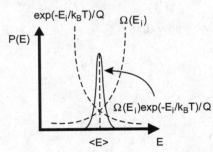

where the sum runs over all microscopic states, or

$$Q(N, V, T) = \sum_i \Omega(N, V, E_i) \exp\left(-\frac{E_i}{k_B T}\right), \qquad (6.18)$$

where the sum runs over energy levels.

Because of Eq. 6.16, the partition function reduces to

$$Q(N, V, T) = \Omega(N, V, \langle E \rangle) \exp\left(-\frac{\langle E \rangle}{k_B T}\right). \qquad (6.19)$$

Taking the natural logarithm of both sides yields

$$\ln Q(N, V, T) = \ln\left(\Omega(N, V, \langle E \rangle) \exp\left(-\frac{\langle E \rangle}{k_B T}\right)\right). \qquad (6.20)$$

The Helmholtz free energy is then

$$A(N, V, T) = \langle E \rangle - k_B T \ln \Omega(N, V, \langle E \rangle). \qquad (6.21)$$

Remembering that

$$A(N, V, T) = U - TS \qquad (6.22)$$

yields for the entropy at the thermodynamic limit

$$S(N, V, T) = k_B \ln \Omega(N, V, \langle E \rangle). \qquad (6.23)$$

The equivalence of ensembles becomes apparent at the thermodynamic limit. Regardless of the ensemble constraints, a system at equilibrium will attain one observable value for each thermodynamic state variable.

6.2 Statistical derivation of the *NVT* partition function

It is instructive before discussing other ensembles to present a different statistical derivation of the $N\dot{V}T$ partition function.

Consider an ensemble of NVT systems (Fig. 6.2). Assume the size of the systems to be large enough to satisfy the thermodynamic limit requirement. Assume also that this ensemble is isolated from the rest of the universe and has an overall constant energy $\mathcal{E}$. If all the systems in the ensemble are in thermal contact with each other, their temperatures will be the same at equilibrium, that is $T_1 = T_2 = \cdots = T_\mathcal{N}$.

In principle, the energy of each of the systems can fluctuate, assuming any one of J energy levels, $E_1, E_2, \ldots E_J$, where J is arbitrarily large $(J \to \infty)$.

The size of the ensemble, $\mathcal{N}$, can also be arbitrarily large. In fact, we can choose it so that $\mathcal{N} \gg J$. We can then define a distribution $\mathcal{D}$ of systems at distinct energy levels. Specifically, the number $\mathcal{N}_j$ of systems with energy E_j is called the occupation number. The set of occupation numbers $\mathcal{N}_1, \mathcal{N}_2, \ldots \mathcal{N}_J$ is the distribution $\mathcal{D}$.

The number of ways a specific distribution can be realized is given by the multinomial distribution coefficient

$$W = \frac{\mathcal{N}!}{\Pi_{j=1}^{J} \mathcal{N}_j!}. \tag{6.24}$$

Maximizing W under the constraints of constant total energy,

$$\mathcal{E} = \sum_{j=1}^{J} \mathcal{N}_j E_j, \tag{6.25}$$

and constant number of ensemble members,

$$\mathcal{N} = \sum_{j=1}^{J} \mathcal{N}_j, \tag{6.26}$$

Figure 6.2 An ensemble of NVT systems. Heat exchange is allowed between systems.

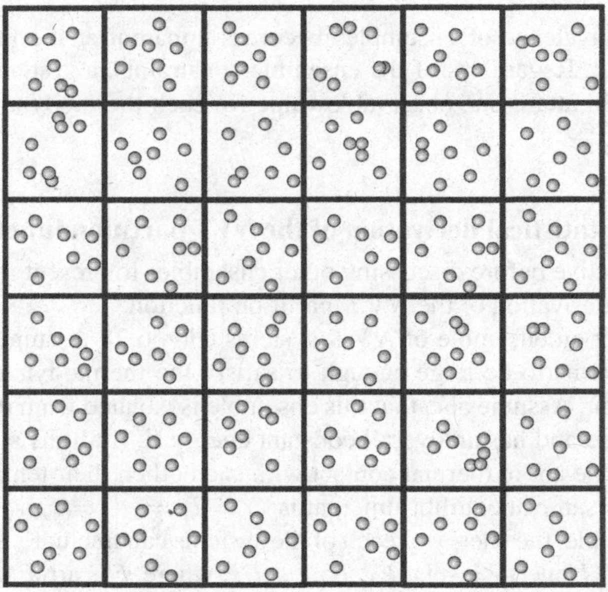

yields the most probable probability distribution $\mathcal{D}$. It can be shown using the method of Lagrange multipliers (this is discussed in optimization textbooks, as well as in advanced statistical thermodynamics texts) that the result of maximizing W is

$$\mathcal{N}_j = c_j \exp(-E_j/k_B T), \tag{6.27}$$

where c_j is a constant. It can actually be shown that the distribution that maximizes W is overwhelmingly more probable than any other distribution. Ultimately, the results obtained from a statistical point of view for the probability distribution and the partition function are the same as the results obtained in Chapter 5.

6.3 Grand-canonical and isothermal-isobaric ensembles

We are now in a position to introduce other ensembles. Two interesting ones for practical applications are the grand-canonical ensemble and the isothermal-isobaric ensemble.

A member system of the grand-canonical ensemble has constant volume and temperature, and it can exchange particles with the environment. At equilibrium, net diffusion of particles stops when the chemical potential of the particles inside the system is the same as outside. The grand-canonical ensemble is also called the $\mu V T$ ensemble (Fig. 6.3). This ensemble is useful for calculating phase equilibria of different systems (see Further reading).

The probability of an open, isothermal system at equilibrium, with constant $\mu V T$, having N_k number of particles and energy E_l can be shown to be

$$P_{lk}(\mu, V, T) = \frac{\exp(-\beta E_l) \exp(-\gamma N_k)}{\Xi}, \tag{6.28}$$

where $\beta = 1/k_B T$ and $\gamma = -\mu/k_B T$.

Note that Ξ is the grand-canonical partition function

$$\Xi(\mu, V, T) = \sum_l \sum_k \exp(-\beta E_l) \exp(-\gamma N_k), \tag{6.29}$$

where the sums run over possible energy levels and number of particles.

We can associate the grand-canonical and the canonical partition functions as follows:

$$\Xi(\mu, V, T) = \sum_k Q(N_k, V, T) \exp(-\gamma N_k). \tag{6.30}$$

The energy of the system in the grand-canonical ensemble is determined to be

$$U = -\frac{\partial}{\partial \beta} \ln \Xi(\mu, V, T). \tag{6.31}$$

Following a similar derivation to the one that yielded the expression relating the Helmholtz free energy to the canonical partition function, we can associate the product PV to the grand-canonical partition function as follows:

$$PV = -k_B T \ln \Xi(\mu, V, T). \tag{6.32}$$

We can define a new variable called the activity, or fugacity, f, expressed as

$$f = \exp\left(-\frac{\mu}{k_B T}\right). \tag{6.33}$$

Figure 6.3 An ensemble of μVT systems.

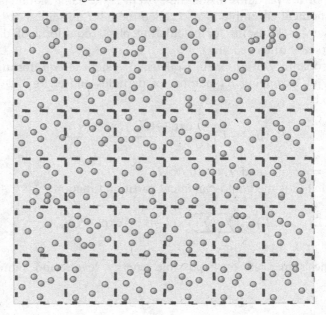

The partition function is then

$$\Xi(\mu, V, T) = \sum_k Q(N_k, V, T) f^{N_k}. \tag{6.34}$$

Using this equation, the average number of atoms in the system is determined to be

$$\langle N \rangle = f \frac{\partial}{\partial f} \ln \Xi(\mu, V, T). \tag{6.35}$$

Similar to the NVT ensemble, fluctuations in the energy vanish at the thermodynamic limit, as do particle number fluctuations. Indeed,

$$\frac{\langle (\delta N) \rangle}{\langle N \rangle} \propto N^{-1/2}. \tag{6.36}$$

In the NPT or isothermal-isobaric ensemble, the pressure and the temperature are kept constant. The NPT ensemble is useful for determining the thermodynamics of systems that are open to the atmosphere, e.g., for biological systems. Following a similar derivation as before, the NPT probability density is

$$\rho_{NPT} = \frac{\exp(-\beta(E + PV))}{\Phi(N, P, T)}, \tag{6.37}$$

where $\Phi(NPT)$ is the isobaric-isothermal ensemble partition function. This is directly related to the Gibbs free energy

$$G = -k_B T \ln \Phi(N, P, T). \tag{6.38}$$

6.4 Maxima and minima at equilibrium

Equation 6.38 indicates that a system at constant temperature and pressure is at equilibrium when its Gibbs free energy is minimum. We have seen that for isolated systems, it is the entropy that is maximum at equilibrium, and that for NVT systems the Helmholtz free energy is minimum at equilibrium.

But how can the entropy for NVE systems change towards a maximum, since according to the Gibbs phase rule for single-component, single-phase systems all thermodynamic property values, or state function values, including the value of entropy, are uniquely determined when the volume and the energy are fixed? Analogous questions can be posed for the NVT and the NPT ensembles.

Answering this question becomes possible by first noting that the entropy is not a property of a single microscopic state. It is instead a property of an ensemble of microscopic states with the same energy, mass, and size constraints. Indeed, any point in microcanonical phase space represents a system with a specific energy value (as well as specific mass, volume, temperature, pressure, and enthalpy values), but not a specific entropy or free energy value. Analogously, any point in canonical phase space represents a system with a specific temperature value (and again with specific energy, mass, volume, pressure, and enthalpy values), but not a specific entropy or free energy value.

Consider, for example, two distinct microscopic states of two separate, isolated systems with N ideal gas particles, energy E and volume V depicted in Fig. 6.4. In the first state, particles are uniformly distributed across the system volume. In the second state, particles are confined in one half of the available system volume. Both of these microscopic states correspond to the same NVE macroscopic state. Both of these microscopic states are available and actually have the same probability of occurring, which is equal to $1/\Omega(N, V, E)$.

This sounds counterintuitive. Indeed, it is not expected that the air in a room will spontaneously move into one half of the room, leaving vacuum in the other half. But this does not mean that this phenomenon is absolutely impossible. It is just highly improbable. The concept of

Figure 6.4 Two systems with identical NVE properties.

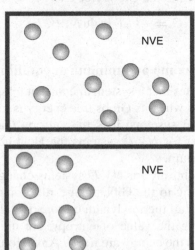

entropy helps us determine exactly how much less probable this strange occurrence is.

So how many NVE ideal gas systems have their particles confined in half the volume? The answer from Eq. 4.61 is

$$\Omega_{IG}(N, V/2, E) = \frac{(V/2)^N}{h^{3N} N!} S_{3N}(2mE)^{3N/2}. \qquad (6.39)$$

On the other hand, how many NVE ideal gas systems have their particles spread over the entire volume? The answer is

$$\Omega_{IG}(N, V, E) = \frac{(V)^N}{h^{3N} N!} S_{3N}(2mE)^{3N/2}. \qquad (6.40)$$

The ratio of the number of available microscopic states occupying half the volume over the number of available microscopic states in the entire volume is simply $(1/2)^N$. At the thermodynamic limit this number goes to zero. Indeed, any number less than one when raised to a power of N_A goes to zero. This discussion can then be extended to any fraction of the volume.

Although each distinct microscopic state in either half or the entire volume has the same probability determined by NVE constraints, the probability that a microscopic state will be observed with the particles in just half the volume is 0.5^N smaller than the probability that a microscopic state will be observed with the particles spread in the entire volume V.

This argument is best captured with the help of the entropy, which is a measure of the number of available microscopic states for distinctly observable macroscopic states. The entropy of an NVE system is $S = k_B \ln \Omega(N, V, E)$ and attains its maximum value at equilibrium, simply because the observable macroscopic state that affords the highest number of microscopic states is overwhelmingly more probable than any other.

Similar arguments can be had for the free energies in μVT and NPT ensembles. That they attain a minimum value at equilibrium is a manifestation of the simple fact that equilibrium systems are observed overwhelmingly more frequently in states that eliminate internal gradients in temperature, pressure, and chemical potential. It is in other words the irrefutable power of statistics that defines equilibria, and indeed much of thermodynamics.

It should be noted that this discussion is no longer valid away from the thermodynamic limit, when the number of particles is small. Fluctuations in small systems can be so large that deviations from the expected thermodynamic behavior are observable, or even dominating.

6.5 Reversibility and the second law of thermodynamics

The previous discussion casts the concept of reversible and irreversible processes in a new light. Consider an isolated system (no mass or energy exchange with its surrounding environment is allowed) with N ideal gas particles, energy E, in volume $V/2$ (Fig. 6.5). The system is confined in $V/2$ by a barrier wall and is at equilibrium. The entropy of this macroscopic state is

$$S_1 = k_B \ln \Omega_{IG}(N, V/2, E) \tag{6.41}$$

or

$$S_1 = k_B \ln \left(\frac{(V/2)^N}{h^{3N} N!} S_{3N} (2mE)^{3N/2} \right). \tag{6.42}$$

If the wall is removed, the gas will expand to volume V. At the new equilibrium state, the entropy is

$$S_2 = k_B \ln \Omega_{IG}(N, V, E) \tag{6.43}$$

Figure 6.5 An irreversible process: expansion from volume $V/2$ to V.

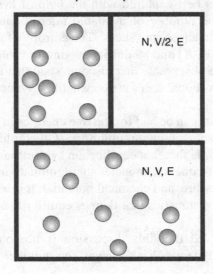

or

$$S_2 = k_B \ln \left(\frac{(V)^N}{h^{3N} N!} S_{3N} (2mE)^{3N/2} \right). \tag{6.44}$$

During the process, the entropy has increased by

$$\Delta S = S_2 - S_1 = k_B \ln \left(\frac{\Omega_{IG}(N, V, E)}{\Omega_{IG}(N, V/2, E)} \right), \tag{6.45}$$

and, finally

$$\Delta S = N k_B \ln 2 > 0. \tag{6.46}$$

This is a well-known result in classical thermodynamics and is in accord with the second law of thermodynamics, which postulates that macroscopic phenomena move in the direction of increasing entropy. This is true not because it is theoretically impossible for the process to be reversed, but because it is practically impossible to observe a process reversal for systems with many particles. In other words, there is a non-zero probability that the system may be observed with all particles confined in $V/2$, but in actuality this probability is so small that we cannot expect to observe a system spontaneously decreasing its entropy.

On the other hand, for a system with, say, $N = 3$ particles, there is a measurable probability that the particles will be confined in half the volume (see Exercises).

This discussion can be extended to NVT systems. Consider the process of free expansion for a system of N particles from volume $V/2$ to volume V. In this case assume that the entire system is in thermal contact with a heat bath of constant temperature. The change in the Helmholtz free energies between the two equilibrium states can be computed with the help of the partition function

$$\Delta A = A_2 - A_1 = k_B \ln \left(\frac{Q(N, V, T)}{Q(N, V/2, T)} \right), \tag{6.47}$$

which quickly yields

$$\Delta A = -N k_B T \ln 2 < 0. \tag{6.48}$$

The Helmholtz free energy of systems at constant volume and temperature changes reaching a minimum at equilibrium.

We can write

$$A = U - TS = E - TS, \tag{6.49}$$

and solving for the entropy

$$S = (E - A)/T. \tag{6.50}$$

For this process then

$$\Delta S = S_2 - S_1 = N k_B \ln 2 > 0 \tag{6.51}$$

as expected.

6.6 Further reading

1. D. A. McQuarrie, *Statistical Mechanics*, (Sausalito, CA: University Science Books, 2000).
2. T. L. Hill, *Statistical Thermodynamics*, (Reading, MA: Addison-Wesley, 1960).
3. F. Reif, *Statistical and Thermal Physics*, (New York: McGraw-Hill, 1965).
4. R. C. Tolman, *Statistical Mechanics*, (London: Oxford University Press, 1938).
5. J. E. Mayer and M. G. Mayer, *Statistical Mechanics*, (New York: Wiley, 1940).
6. D. Chandler, *Introduction to Modern Statistical Mechanics*, (New York: Oxford University Press, 1987).
7. T. L. Hill, *Statistical Mechanics: Principles and Selected Applications*, (New York: Dover, 1987).

6.7 Exercises

1. Derive an expression for the partition function $\Delta(N, P, T)$ of the isothermal-isobaric ensemble.

2. Prove that the Gibbs free energy, $G(N, P, T)$, is given in terms of the partition function for the isothermal-isobaric ensemble by

$$G(N, P, T) = -\frac{1}{\beta} \ln \Delta(N, P, T).$$

3. Prove that in the grand-canonical ensemble the mean-square fluctuation of the number of particles can be expressed as

$$\langle (\delta N)^2 \rangle = \langle N^2 \rangle - \langle N \rangle^2 = f \frac{\partial}{\partial f} \left(f \frac{\partial}{\partial f} \ln \Xi(\mu, V, T) \right).$$

4. Calculate the volume fluctuations in the NPT ensemble. Express your answer in terms of the isothermal compressibility, κ, defined as

$$\kappa = -\frac{1}{V} \left. \frac{\partial V}{\partial P} \right|_{N,T}.$$

Show that fluctuations vanish in the thermodynamic limit. What happens at the critical point?

5. Consider a two-dimensional membrane system. In the thermodynamic description of such systems, the area (A) plays the role that volume (V) does in three-dimensional systems, and the surface pressure (π) replaces the pressure P. Analogously to three-dimensional systems, the work exerted in increasing the area is (πdA). The fundamental thermodynamic equation for the energy is thus

$$dE = TdS - \pi dA.$$

The fundamental thermodynamic equation for the Helmholtz free energy ($F = E - TS$) is

$$dF = -SdT - \pi dA.$$

The energy levels of a free particle in a two-dimensional periodic box are determined as

$$E(n_x, n_y) = \hbar^2 \frac{4\pi^2}{2m} \left(\frac{n_x^2}{L_x^2} + \frac{n_y^2}{L_y^2} \right),$$

where $-\infty \leq n_x, n_y \leq \infty$, and $A = L_x L_y$.

Use this result to calculate the partition function, the energy, the film pressure, and the entropy of a two-dimensional perfect gas.

6. Start with $\langle P \rangle Q$ for an NVT ensemble and verify that the relative fluctuations in pressure are

$$\frac{\langle (\delta P) \rangle}{\langle (P) \rangle} = k_B T \left[\left(\frac{\partial \langle P \rangle}{\partial V} \right)_{N,T} - \left\langle \frac{\partial^2 E}{\partial V^2} \right\rangle_{N,T} \right].$$

Comment on the relative importance of pressure fluctuations for an ideal gas.

7. How many molecules of an ideal gas do you expect to find in a volume of $1000\ \text{nm}^3$ at 1 atm and 300 K? How confident are you?

8. Consider an NVT system with $N = 3$. What is the probability that all particles are confined in half the system volume?

7

Molecules

Thus far, we have only introduced systems of monoatomic, or point mass, particles. In this chapter we present a statistical mechanical derivation of polyatomic system thermodynamics. We focus on diatomic molecules in an ideal gas phase and we use the canonical ensemble partition function. The reason this ensemble is chosen is the ease with which the integration of the Boltzmann factor can be performed over the entire phase space. All the ensembles of molecular systems are again equivalent at the thermodynamic limit.

7.1 Molecular degrees of freedom

Consider a system of N non-interacting molecules at constant temperature T in volume V. Generally, for a molecule of s atoms, $3s$ positions and $3s$ conjugate momenta are needed to determine uniquely the molecular microscopic state. Note that we are still neglecting subatomic degrees of freedom, such as electronic and nuclear degrees of freedom, assuming each atom to be a point mass. This assumption is valid and does not influence the accuracy of results for systems at high temperatures (room temperature is typically high enough).

Each molecule i has a Hamiltonian $h_i(\underline{p}_i, \underline{q}_i)$, where $\underline{p}_i$ and $\underline{q}_i$ are each a $3s$-dimensional vector. In the absence of intermolecular forces, the total system Hamiltonian is

$$H(\underline{p}, \underline{q}) = \sum_{i=1}^{N} h_i(\underline{p}_i, \underline{q}_i), \tag{7.1}$$

where $\underline{p}$ and $\underline{q}$ are now each a $(3Ns)$-dimensional vector.

The $\overline{NVT}$ partition function is

$$Q(N, V, T) = \frac{1}{N! \, h^{3Ns}} \int d\underline{p} d\underline{q} \, \exp\left(-\beta \sum_{i=1}^{N} h_i\right) \tag{7.2}$$

or

$$Q(N, V, T) = \frac{1}{N! \, h^{3Ns}} \prod_{i=1}^{N} \int d\underline{p}_i \, d\underline{q}_i \, \exp(-\beta h_i). \qquad (7.3)$$

The integral is the same for all molecules in the system. Consequently, for any molecular system

$$Q(N, V, T) = \frac{q(1, V, T)^N}{N!}, \qquad (7.4)$$

where

$$q(1, V, T) = \frac{1}{h^{3s}} d\underline{p}_i \, d\underline{q}_i \, \exp(-\beta h_i) \qquad (7.5)$$

is defined as the molecular partition function.

It is apparent then that thermodynamic state functions for molecules depend on the functional form of the molecular Hamiltonian and its dependence on the molecular degrees of freedom. To see exactly how, let us consider diatomic molecules in the ideal gas phase.

7.2 Diatomic molecules

Consider a system of N diatomic molecules in the NVT ensemble. The microscopic state and the partition function of each molecule can be defined if the positions and momenta of the two atoms are known (Fig. 7.1). There are thus 12 degrees of freedom for each molecule, six

Figure 7.1 A diatomic molecule.

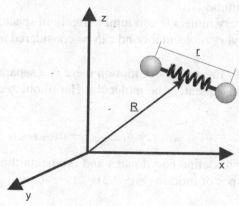

positions and six conjugate momenta. As a result, the molecular phase space of the entire NVT system has 12 dimensions.

Using a Cartesian coordinate system, the microscopic state of each molecule i could be defined by the vector

$$\underline{X}_i = \underline{X}_i(x_1, \; y_1, \; z_1, \; x_2, \; y_2, \; z_2, \; px_1, \; py_1, \; pz_1, \; px_2, \; py_2, \; pz_2),$$
$$(7.6)$$

or, more compactly

$$\underline{X} = \underline{X}(\underline{r}_1, \; \underline{r}_2, \; \underline{p}_1, \; \underline{p}_2),   \tag{7.7}$$

where $\underline{r}_j = (x_j, y_j, z_j)$ are the coordinates of atom j and $\underline{p}_j = (px_j, \; py_j, \; pz_j)$ are the conjugate momenta.

The Hamiltonian of the molecule can be written as

$$h_i = \frac{p_1^2}{2m_1} + \frac{p_2^2}{2m_2} + U(|\underline{r}_1 - \underline{r}_2|),   \tag{7.8}$$

where U is the potential of the bond linking the two atoms (more about this later in this chapter).

Although conceptually straightforward, the use of Cartesian coordinates and momenta is not convenient when describing the motion of a diatomic molecule. This is because these degrees of freedom do not change independently, since the chemical bond between the two atoms constrains their motion.

A convenient classical mechanical model of a diatomic molecule includes the following types of motion, assumed, at a first approximation, as independent:

1. Translational; every molecule is free to move, in terms of its center of mass, in volume V;
2. Rotational; every molecule can tumble freely in space;
3. Vibrational; every molecular bond can be considered as a spring that vibrates.

For each one of these types of motion there is a separate term in the Hamiltonian. Consequently, the molecular Hamiltonian can be written as

$$h_i = h_{\text{translation}, i} + h_{\text{rotation}, i} + h_{\text{vibration}, i}.   \tag{7.9}$$

It is then best to define coordinates and momenta that can describe these separate types of motion (Fig. 7.2).

The generalized, or internal, coordinates which better describe the orientation, shape, and motion of linear diatomic molecules are the following:

1. Three coordinates $\underline{R}$ are needed to define the position of the center of mass of the molecule. The three conjugate momenta $\underline{P}$ describe the translational motion of the molecule in space. These can as well be defined in a Cartesian coordinate system, i.e., have $\underline{R} \equiv (x, y, z)$ and $\underline{P} \equiv (P_x, P_y, P_z)$.

2. Two angles, ϕ and θ, are needed to define the orientation of a linear diatomic molecule. The two conjugate momenta, p_ϕ and p_θ, describe the rotational motion of the molecule.

3. One distance, r, is needed to define the size of the bond and a conjugate momentum, p_r, can define the vibrational motion of the molecule.

These generalized coordinates are equivalent to the 12 Cartesian coordinates and each set can be derived from the other with coordinate transformation relationships, as discussed in Chapter 3.

For simplicity, we ignore the interaction between these separate degrees of freedom. For example, at low temperatures, the rotational

Figure 7.2 Types of motion of diatomic molecules.

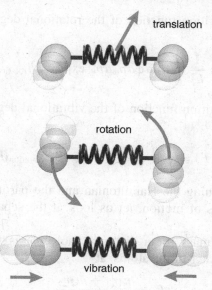

motion may influence the vibrational motion of a diatomic molecule, but at high temperatures we can neglect such coupling between degrees of freedom.

The Hamiltonian for each molecule i can now be written as

$$h_i(\underline{R}, \underline{P}, \phi, \theta, p_\phi, p_\theta, r, p_r)$$
$$= h_{\text{translation}}(\underline{R}, \underline{P}) + h_{\text{rotation}}(\phi, \theta, p_\phi, p_\theta) + h_{\text{vibration}}(r, p_r). \tag{7.10}$$

Consequently, the partition function for one molecule can be written as

$$q(1, V, T) = q_{\text{translation}}(1, V, T) \, q_{\text{rotation}}(1, V, T) \, q_{\text{vibration}}(1, V, T) \tag{7.11}$$

or

$$q(1, V, T) = q_{\text{translation}}(1, V, T) \, q_{\text{internal}}(1, V, T), \tag{7.12}$$

where q_{internal} is the partition function of the internal degrees of freedom of a molecule, which include rotations and vibrations.

The partition function of the translational degrees of freedom is

$$q_{\text{translation}}(1, V, T) = \frac{1}{h^3} \int d\underline{R} d\underline{P} \, \exp\left(-\beta h_{\text{translation}}(\underline{R}, \underline{P})\right). \tag{7.13}$$

In turn, the partition function of the rotational degrees of freedom is

$$q_{\text{rotation}}(1, V, T) = \frac{1}{h^2} \int d\theta d\phi dp_\phi dp_\theta \, \exp\left(-\beta h_{\text{rotation}}(\phi, \theta, p_\phi, p_\theta)\right). \tag{7.14}$$

Finally, the partition function of the vibrational degrees of freedom is

$$q_{\text{vibration}}(1, V, T) = \frac{1}{h} \int dr \, dp_r \, \exp\left(-\beta h_{\text{vibration}}(r, p_r)\right). \tag{7.15}$$

Before determining the Hamiltonian and the partition function for the different types of motion, let us look at the separate terms more carefully.

The center of mass for a diatomic molecule is found, using the Cartesian coordinates of the atoms, at

$$\underline{R} = \frac{m_1 \underline{r}_1 + m_2 \underline{r}_2}{m_1 + m_2}, \tag{7.16}$$

where m_1 and m_2 are the masses of the two atoms and $\underline{r}_1$ and $\underline{r}_2$ are the atom positions. If the masses of the two atoms are the same, then

$$\underline{R} = \frac{\underline{r}_1 + \underline{r}_2}{2}. \tag{7.17}$$

The center of mass momentum vector is

$$\underline{P} = \underline{P}_1 + \underline{P}_2. \tag{7.18}$$

The bond length is

$$r = |\underline{r}_1 - \underline{r}_2|. \tag{7.19}$$

The conjugate bond-length momentum vector is

$$\underline{p} = (p_r, p_\phi, p_\theta) = \frac{m_2\,\underline{P}_1 - m_1\,\underline{P}_2}{m_1 + m_2}. \tag{7.20}$$

The Hamiltonian of the molecule, written in terms of internal coordinates, is then

$$h_i = \frac{P^2}{2M} + \frac{p^2}{2\mu} + U(r), \tag{7.21}$$

where

$$M = m_1 + m_2 \tag{7.22}$$

is the total mass of the molecule and

$$\mu = \frac{m_1 m_2}{m_1 + m_2} \tag{7.23}$$

is defined as the reduced mass of the molecule.

The molecular Hamiltonian can be expanded further as

$$h_i = \frac{P^2}{2M} + \frac{1}{2\mu}\left(p_r^2 + \frac{p_\theta^2}{r^2} + \frac{p_\phi^2}{r^2 \sin^2\theta}\right) + U(r). \tag{7.24}$$

In all cases the integral over the translational degrees of freedom in Eq. 7.13 can be carried out exactly:

$$q_{\text{translation}}(1, V, T) = \frac{1}{h^3} \int d\underline{R}\,d\underline{P}\, \exp\left(-\beta\frac{P^2}{2M}\right). \tag{7.25}$$

Finally,

$$q_{\text{translation}}(1, V, T) = V\left(\frac{2\pi M k_B T}{h^2}\right)^{3/2}. \tag{7.26}$$

This result can be generalized to any molecule. Indeed, the translational partition function is always computed with Eq. 7.26, where M is the total molecular mass.

Depending on the temperature, the internal degrees of freedom of rotation and vibration differ in their relative contribution to the total number of available microscopic states and the value of the partition function. We present two cases with varying relative contributions by the rotational and vibrational motions.

7.2.1 Rigid rotation

Figure 7.3 illustrates the bond potential $U(r)$ of a diatomic molecule. By definition, there is a zero energy, equilibrium bond length, r_o. If the separation of the two atoms is less than the equilibrium length, the two atoms experience a repulsive force. The interaction strength becomes infinite at very small distances, representing the excluded volume effect. On the other hand, if the distance is larger than the equilibrium length, the atoms experience an attractive force. The bond breaks, however, if the distance becomes larger than a certain value, which depends on the nature of the atoms.

If the minimum in U is sufficiently sharp, the distance between the atoms is virtually constrained, with $r \approx r_o$. The vibrational motion can then be neglected. Consequently, the molecule is considered to tumble like a dumbbell, or rigid rotor, and the vibrational partition function is simply

$$q_{\text{vibration}} = 1. \tag{7.27}$$

Figure 7.3 Bond potential $U(r)$ of a diatomic molecule.

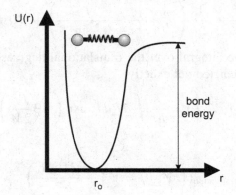

The internal Hamiltonian of a rigid rotor is written as

$$h_{\text{internal}} = \frac{1}{2\mu} \left(\frac{p_\theta^2}{r_o^2} + \frac{p_\phi^2}{r_o^2 \sin^2 \theta} \right). \tag{7.28}$$

The rotational partition function is then determined as

$$q_{\text{rotation}}(1, V, T)$$
$$= \frac{1}{h^2} \int d\theta d\phi dp_\phi dp_\theta \exp \left(-\beta \left[\frac{1}{2\mu r_o^2} \left(p_\theta^2 + \frac{p_\phi^2}{\sin^2 \theta} \right) \right] \right). \tag{7.29}$$

We can integrate p_ϕ and p_θ in the interval $[-\infty, +\infty]$, θ in the interval $[0 - \pi]$, and ϕ in the interval $[0 - 2\pi]$ to obtain

$$q_{\text{rotation}}(1, V, T) = \frac{8\pi^2 \mu r_o^2}{\beta h^2} \tag{7.30}$$

or

$$q_{\text{rotation}}(1, V, T) = \frac{8\pi^2 \mu r_o^2 k_B T}{h^2}. \tag{7.31}$$

Consequently, in the limiting case of negligible vibrations, the NVT ensemble partition function of an ideal gas of diatomic molecules is

$$\ln Q(N, V, T) = N \ln q - \ln N!$$
$$= N(\ln q_{\text{translation}} + \ln q_{\text{rotation}}) - \ln N!$$
$$= N \ln \left(V \left(\frac{2\pi M k_B T}{h^2} \right)^{3/2} \times \left(\frac{8\pi^2 \mu r_o^2 k_B T}{h^2} \right) \right) - \ln N!, \tag{7.32}$$

Simplifying yields for the rigid rotor partition function

$$\ln Q(N, V, T) = N \left(\ln V - \frac{5}{2} \ln \beta + C \right), \tag{7.33}$$

where C is a system constant.

Following the analysis of previous chapters, all thermodynamic properties can now be derived for diatomic ideal gases.

Of particular interest are the following:

1. The pressure is found again to be

$$P = \frac{N k_B T}{V}. \tag{7.34}$$

This is a well-known result, that the equation of state for ideal gases is not dependent on molecular structure.

2. The energy is found to be

$$E = \frac{5}{2} N k_B T. \tag{7.35}$$

Interestingly, the energy of diatomic molecules with free rotation is higher than the energy of monoatomic molecules at the same temperature. We will see later in this chapter that each quadratic degree of freedom in the Hamiltonian adds $\frac{1}{2} k_B T$ per molecule to the energy of the system. For rigid rotors, the two rotational degrees of freedom, p_θ and p_ϕ, together add $k_B T$ per molecule.

3. The constant volume heat capacity is

$$C_V = \frac{5}{2} N k_B. \tag{7.36}$$

This is larger than the monoatomic particle heat capacity by k_B per molecule. This makes sense: because of the additional internal degrees of freedom, more energy is needed to increase the temperature of the system by one degree.

7.2.2 Vibrations included

Consider a system of diatomic molecules at a temperature where vibrations can no longer be neglected.

The vibrational Hamiltonian is

$$h_{\text{vibration}} = \frac{1}{2\mu}(p_r^2 + U(r)). \tag{7.37}$$

According to Eq. 7.15, the vibrational partition function is

$$q_{\text{vibration}}(1, V, T) = \frac{1}{h^2} \int dp_r \, \exp\left(-\frac{\beta}{2\mu} p_r^2\right) \int dr \, \exp(-\beta U(r)). \tag{7.38}$$

The first integral is straightforward to compute, since there is a quadratic term in the exponential:

$$\int dp_r \exp\left(-\frac{\beta}{2\mu} p_r^2\right) = \left(\frac{8\pi\mu}{\beta}\right)^{1/2}. \tag{7.39}$$

The second integral is computed with limits $[0, +\infty]$. The partition function is then

$$q_{\text{vibration}}(1, V, T) = \left(\frac{8\pi\mu}{\beta h^2}\right)^{1/2} \int_0^{+\infty} dr \, \exp\left(-\beta U(r)\right). \tag{7.40}$$

At the equilibrium length r_o the bond energy is $U(r_o) = 0$. A deviation length can be defined from the equilibrium as $\delta = r - r_o$.

A reasonable approximation for the bond energy potential is the quadratic one, the result of a Taylor expansion around r_o for small deviations from the equilibrium bond length. This is expressed as follows:

$$U(r) = U(r_o + \delta) \approx U(r_o) + \delta \left.\frac{\partial U}{\partial r}\right|_{r_o} + \frac{1}{2}\delta^2 \left.\frac{\partial^2 U}{\partial r^2}\right|_{r_o}. \qquad (7.41)$$

By definition, the first two terms in the sum are zero. The value of the second derivative of the potential with respect to the bond length is a constant at the equilibrium length, that is

$$\left.\frac{\partial^2 U}{\partial r^2}\right|_{r_o} = k. \qquad (7.42)$$

The vibrational motion of a diatomic molecule can then be approximated with a spring of constant $k = \mu\omega^2$, where ω is the characteristic oscillatory frequency of vibrations.

Then, and since $dr = d\delta$, we can write

$$\int_0^{+\infty} dr \, \exp(-\beta U(r)) = \int_{-\infty}^{+\infty} d\delta \, \exp\left(-\frac{\beta\mu\omega^2}{2}\delta^2\right)$$

$$= \left(\frac{\pi}{2\beta\mu\omega^2}\right). \qquad (7.43)$$

Consequently, the vibrational partition function is

$$q_{\text{vibration}}(1, V, T) = \frac{2\pi k_B T}{h\omega}. \qquad (7.44)$$

Again, the single molecule partition function is

$$q(1, V, T) = q_{\text{translation}}(1, V, T) \, q_{\text{rotation}}(1, V, T) \, q_{\text{vibration}}(1, V, T). \qquad (7.45)$$

Combining all the molecular partition functions we find

$$q(1, V, T) = V \left(\frac{2\pi M k_B T}{h^2}\right)^{3/2} \left(\frac{8\pi^2 \mu r_o^2 k_B T}{h^2}\right) \left(\frac{2\pi k_B T}{h\omega}\right). \qquad (7.46)$$

Simplifying yields for the entire N-molecule system partition function

$$\ln Q(N, V, T) = N \left(\ln V - \frac{7}{2}\ln\beta + C\right), \qquad (7.47)$$

where C is a system constant.

Following the same analysis as in the rigid rotor case, we compute the following thermodynamic properties:

1. The pressure, as expected, is found again to be

$$P = \frac{Nk_BT}{V}.$$ (7.48)

2. The energy is now

$$E = \frac{7}{2}Nk_BT.$$ (7.49)

If we include the vibrational motion, there are two additional quadratic degrees of freedom in the Hamiltonian (for δ and p_r) adding k_BT per molecule to the energy of the system.

3. The constant volume heat capacity is now

$$C_V = \frac{7}{2}Nk_B.$$ (7.50)

The temperature of the system determines the validity of each of the approximations made that impact the value of the heat capacity of diatomic molecules. At high temperatures, vibrations are important and the heat capacity is best described by Eq. 7.50. For lower temperatures, the rigid rotor approximation becomes valid and Eq. 7.36 captures the heat capacity. The temperature thresholds depend on the molecule type. In Fig. 7.4, the heat capacity of a diatomic molecule is shown as a function of temperature.

A more accurate treatment to determine the valid temperature ranges for various heat capacity values must include quantum mechanical arguments, which are beyond the scope of this book. The specific heat results of a quantum mechanical treatment of a diatomic molecule are

Figure 7.4 Constant volume specific heat of a diatomic molecule.

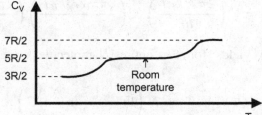

summarized as follows:

$$C_V = \begin{cases} \dfrac{5}{2} N k_B T, & \text{if } \hbar\omega \gg k_B T \\[2mm] \dfrac{7}{2} N k_B T, & \text{if } \hbar\omega \ll k_B T \end{cases} \qquad (7.51)$$

where

$$\hbar = \frac{h}{2\pi} \qquad (7.52)$$

and ω is the characteristic angular frequency of bond vibrations.

7.2.3 Subatomic degrees of freedom

The discussion in the preceding section is premised on the assumption that subatomic degrees of freedom are not important for accurate calculation of thermodynamic properties. Naturally, these subatomic degrees of freedom are associated with the state of electrons and nuclei in atoms. In principle, Eq. 7.10 should be modified to include the nuclear and electronic Hamiltonian

$$h_i = h_{\text{translation},i} + h_{\text{rotation},i} + h_{\text{vibration},i} + h_{\text{nuclear},i} + h_{\text{electronic},i}.$$
$$(7.53)$$

The partition function for one molecule can then be written as

$$q = q_{\text{translation}}\, q_{\text{rotation}}\, q_{\text{vibration}}\, q_{\text{nuclear}}\, q_{\text{electronic}}. \qquad (7.54)$$

Of course, as we have discussed, the nuclear and electronic degrees of freedom do not contribute to the molecular partition function at ordinary temperatures. For the nuclear term, this is because the two lowest energy levels of the nucleus are separated by an energy of $O(1\,\text{MeV})$. This means that only at temperatures of $O(10^{10})\,$K would $k_B T \approx 1\,\text{MeV}$, rendering the nuclear degrees of freedom important. For electrons, the lowest energy levels are only separated by approximately $1\,\text{eV}$. This means that for temperatures less than approximately $10\,000\,$K the atom would be in its electronic ground state and the electronic partition function can be neglected. We can then safely assume that $q_{\text{nuclear}} = 1$ and $q_{\text{electronic}} = 1$.

7.3 Equipartition theorem

At this point it is useful to prove the equipartition theorem. This theorem states that for each quadratic degree of freedom in the Hamiltonian, a term equal to $1/(2\beta)$ is added to the energy. This is a particularly useful theorem in statistical mechanics that nicely explains how the heat capacity of matter changes with temperature.

Consider a system with n distinct, variable degrees of freedom, denoted by a vector $\underline{\xi}$. The Hamiltonian is defined as

$$H = H(\underline{\xi}) = H(\xi_1, \ldots, \xi_n). \tag{7.55}$$

These degrees of freedom could be the positions and momenta of particles. Consider the kth degree of freedom and assume that the Hamiltonian can be split additively as follows:

$$H = h(\xi_k) + H'(\xi_1, \ldots, \xi_{k-1}, \xi_{k+1}, \ldots, \xi_n). \tag{7.56}$$

Further assume that for this kth degree of freedom, the Hamiltonian term is quadratic in ξ_k, i.e.,

$$h(\xi_k) = \lambda \xi_k^2 = \varepsilon_k, \tag{7.57}$$

where λ is an arbitrary factor and ε_k is the energy contribution of the ξ_k degree of freedom.

The ensemble average value of ε_k is (neglecting any correction prefactors)

$$<\varepsilon_k> = \frac{\int \varepsilon_k \exp(-\beta H(\underline{\xi}))d\underline{\xi}}{\int \exp(-\beta H(\underline{\xi}))d\underline{\xi}}. \tag{7.58}$$

Using Eq. 7.56

$$<\varepsilon_k> = \frac{\int \varepsilon_k \exp(-\beta \varepsilon_k) \exp(-\beta H'(\xi_1, \ldots, \xi_{k-1}, \xi_{k+1}, \ldots \xi_n))d\underline{\xi}}{\int \exp(-\beta \varepsilon_k) \exp(-\beta H'(\xi_1, \ldots, \xi_{k-1}, \xi_{k+1}, \ldots \xi_n))d\underline{\xi}}. \tag{7.59}$$

Since H' is not a function of ξ_k we can simplify as follows:

$$<\varepsilon_k> = $$
$$\frac{\int \varepsilon_k \exp(-\beta \varepsilon_k)d\xi_k \int \exp(-\beta H'(\xi_1, \ldots, \xi_{k-1}, \xi_{k+1}, \ldots \xi_n))d\xi_1 \ldots d\xi_{k-1} d\xi_{k+1} \ldots d\xi_n}{\int \exp(-\beta \varepsilon_k)d\xi_k \int \exp(-\beta H'(\xi_1, \ldots, \xi_{k-1}, \xi_{k+1}, \ldots \xi_n))d\xi_1 \ldots d\xi_{k-1} d\xi_{k+1} \ldots d\xi_n}, \tag{7.60}$$

which yields

$$<\varepsilon_k> = \frac{\int \varepsilon_k \exp(-\beta \varepsilon_k)d\xi_k}{\int \exp(-\beta \varepsilon_k)d\xi_k}. \tag{7.61}$$

We can then write

$$\int \varepsilon_k \exp(-\beta \varepsilon_k)d\xi_k = -\frac{\partial}{\partial \beta} \int \exp(-\beta \varepsilon_k)d\xi_k. \tag{7.62}$$

Consequently,

$$< \varepsilon_k > = -\frac{\partial}{\partial \beta} \ln \left(\int \exp(-\beta \varepsilon_k) d\xi_k \right). \tag{7.63}$$

Using Eq. 7.57

$$\int \exp(-\beta \varepsilon_k) d\xi_k = \int \exp(-\beta \lambda \xi_k^2) d\xi_k. \tag{7.64}$$

Since the limits of the integration extend to infinity, we can introduce a new variable $t = \sqrt{\lambda \beta} \xi_k$ to write

$$\int \exp(-\beta \lambda \xi_k^2) d\xi_k = \frac{1}{\sqrt{\lambda \beta}} \int \exp(-t^2) dt = \sqrt{\frac{\pi}{\lambda \beta}}. \tag{7.65}$$

Returning to Eq. 7.63, a few simple algebraic operations yield

$$< \varepsilon_k > = \frac{1}{2\beta} = \frac{k_B T}{2}, \tag{7.66}$$

which proves the equipartition theorem.

In the following chapter we derive the virial theorem, which is related to the equipartition theorem. The virial theorem is a starting point for a discussion of non-ideal gases.

7.4 Further reading

1. D. A. McQuarrie and J. D. Simon, *Molecular Thermodynamics*, (Sausalito, CA: University Science Books, 1999).
2. H. Metiu, *Physical Chemistry, Statistical Mechanics*, (New York: Taylor and Francis, 2006).
3. D. Chandler, *Introduction to Modern Statistical Mechanics*, (New York: Oxford University Press, 1987).
4. T. L. Hill, *Statistical Mechanics: Principles and Selected Applications*, (New York: Dover, 1987).
5. D. A. McQuarrie, *Statistical Mechanics*, (Sausalito, CA: University Science Books, 2000).

7.5 Exercises

1. Prove that Eq. 7.20 is true.

2. In quantum mechanics, the energy levels of a rigid rotor are

$$E_{\text{rot}, j} = \frac{j(j+1)h^2}{8\pi I},$$

where $j = 0, 1, 2, \ldots$, and $I = \mu r_o^2$ is the moment of inertia about the center of mass. Find the partition function in terms of the characteristic temperature of rotation

$$\Theta_r = \frac{h^2}{8\pi k_B I}.$$

What is the contribution of rotations to the internal energy and constant volume heat capacity?

3. The moment of inertia for oxygen is $I = 1.9373 \times 10^{-46}\,\text{kg m}^2$ at $T = 298.15$ K. What is the rotational contribution to the constant volume heat capacity, the molar entropy, and the molar enthalpy?

4. According to quantum mechanics, the energy levels of a vibrating diatomic molecule are

$$E_{\text{vib},n} = \left(n + \frac{1}{2}\right) h\upsilon,$$

where $n = 0, 1, 2, \ldots$, and υ is the characteristic frequency of oscillations. Derive the vibrational motion partition function in terms of the characteristic temperature of vibration

$$\Theta_\upsilon = \frac{h\upsilon}{k_B}.$$

What is the contribution of vibrations to the internal energy and constant volume heat capacity?

5. The characteristic temperature of vibrations for oxygen is $\Theta_\upsilon = 2273$ K at $T = 298$ K. What is the vibrational contribution to the heat capacity? How does it compare to the translational contribution?

6. What is the vibrational frequency of HI? Assume the force constant to be $k = 317\,\text{Nm}^{-1}$.

7. Starting with the kinetic energy, K, of a diatomic molecule in terms of Cartesian coordinates, derive the kinetic energy in terms of spherical coordinates.

8

Non-ideal gases

In this chapter we develop the statistical mechanical theory that is necessary to determine the thermodynamic behavior of non-ideal gases. For non-ideal gases, the potential energy of particle–particle interactions is no longer negligible, as a result of higher densities.

The Hamiltonian for a system of N particles is written as

$$H(\underline{p}, \underline{q}) = K(\underline{p}) + U(\underline{q})$$

$$= \sum_{i=1}^{3N} \frac{p_i^2}{2m} + U(\underline{q}). \tag{8.1}$$

Assuming constant volume V and temperature T, the canonical partition function is again

$$Q(N, V, T) = \frac{1}{N! h^{3N}} \int d\underline{p} \exp\left(-\frac{\beta}{2m} \sum_{i=1}^{3N} p_i^2\right) \int d\underline{q} \exp\left(-\beta U(\underline{q})\right). \tag{8.2}$$

The integration over the momenta is again simple, yielding

$$Q(N, V, T) = \frac{1}{N!} \left(\frac{2\pi m k_B T}{h^2}\right)^{3N/2} \int d\underline{q} \exp\left(-\beta U(\underline{q})\right). \tag{8.3}$$

We define the configurational integral of NVT systems as

$$Z(N, V, T) = \int d\underline{q} \exp\left(-\beta U(\underline{q})\right). \tag{8.4}$$

Equation 8.4 is an exact yet stupendously complicated formula. The configurational integral encompasses all non-ideal behavior, and prescribes the dependence of the partition function and the thermodynamics of materials on the density.

For example, equations of state for non-ideal gases solely depend on the configurational integral. Specifically, the pressure is determined as

$$P = -\frac{\partial A}{\partial V}\bigg|_{N,T}$$

$$= \frac{1}{\beta}\frac{\partial \ln Q}{\partial V}\bigg|_{N,T}$$

$$= k_B T \frac{\partial \ln Z}{\partial V}\bigg|_{N,T}. \tag{8.5}$$

In this chapter, we discuss different types of classical mechanical atomic interaction and forms of the potential energy U in non-ideal gases, for which the configurational integral Z and the thermodynamic properties can be determined exactly. We begin with the virial theorem, which provides a bridge to numerical simulations for computing equations of state for systems whose interactions are not amenable to analytical calculations.

8.1 The virial theorem

Consider a system with N particles in volume V with constant temperature T. There are $6N$ degrees of freedom that define the microscopic states of this system:

generalized coordinates: $q_1, q_2 \ldots, q_{3N}$

generalized momenta: $p_1, p_2 \ldots, p_{3N}.$ \hfill (8.6)

We can define a new set of variables, $\underline{\xi}$, that do not distinguish between coordinates and momenta, as follows:

$$\xi_1 \equiv p_1, \xi_2 \equiv p_2, \ldots, \xi_{3N} \equiv p_{3N},$$
$$\xi_{3N+1} \equiv q_1, \xi_{3N+2} \equiv q_2, \ldots, \xi_{6N} \equiv q_N. \tag{8.7}$$

The Hamiltonian is a function of the $6N$ $\underline{\xi}$ variable degrees of freedom, i.e.,

$$H = H(\xi_1, \ldots, \xi_{6N}). \tag{8.8}$$

A volume element of phase space can be defined as

$$d\underline{X} = \prod_{i=1}^{3N} dp_i dq_i \tag{8.9}$$

or, alternatively, in terms of the new variables, as

$$d\underline{\xi} = \prod_{k=1}^{6N} d\xi_k. \tag{8.10}$$

Let us now consider the mean value of the function $\xi_k(\partial H/\partial \xi_k)$

$$\left\langle \xi_k \frac{\partial H}{\partial \xi_k} \right\rangle = \frac{\int \xi_k \frac{\partial H}{\partial \xi_k} \exp\left(-\beta H(\underline{\xi})\right) d\underline{\xi}}{\int \exp\left(-\beta H(\underline{\xi})\right) d\underline{\xi}}. \tag{8.11}$$

We can write

$$\frac{\partial H}{\partial \xi_k} \exp(-\beta H(\underline{\xi})) = -\frac{1}{\beta} \frac{\partial \exp(-\beta H(\underline{\xi}))}{\partial \xi_k}. \tag{8.12}$$

Equation 8.11 now yields

$$\left\langle \xi_k \frac{\partial H}{\partial \xi_k} \right\rangle = \frac{-\frac{1}{\beta} \int \xi_k \frac{\partial \exp(-\beta H(\underline{\xi}))}{\partial \xi_k} d\underline{\xi}}{\int \exp\left(-\beta H(\underline{\xi})\right) d\underline{\xi}}. \tag{8.13}$$

The numerator can be integrated by parts, using $\exp(-\beta H(\underline{\xi})) \to 0$ at the limits of integration $\xi_k \to -\infty$ and $\xi_k \to +\infty$.

After some algebra,

$$\left\langle \xi_k \frac{\partial H}{\partial \xi_k} \right\rangle = \frac{\frac{1}{\beta} \int \exp\left(-\beta H(\underline{\xi})\right) d\underline{\xi}}{\int \exp\left(-\beta H(\underline{\xi})\right) d\underline{\xi}}, \tag{8.14}$$

which simplifies to

$$\left\langle \xi_k \frac{\partial H}{\partial \xi_k} \right\rangle = \frac{1}{\beta}. \tag{8.15}$$

This is the general mathematical representation of the virial theorem, applicable to all microscopic degrees of freedom.

The virial theorem was first stated by Clausius in 1870 for the expectation value of the product between the position of each particle and the force acting on it. Indeed, substituting in Eq. 8.15 for the position of a particle, and using Hamilton's equation of motion, yields

$$\left\langle q_k \frac{\partial H}{\partial q_k} \right\rangle = k_B T, \quad k = 1, \dots, 3N \tag{8.16}$$

or

$$-\langle q_k \dot{p}_k \rangle = k_B T, \quad k = 1, \ldots, 3N \tag{8.17}$$

and finally

$$-\langle q_k F_k \rangle = k_B T, \quad k = 1, \ldots, 3N. \tag{8.18}$$

8.1.1 Application of the virial theorem: equation of state for non-ideal systems

In vector notation and considering all three coordinates, Eq. 8.16 becomes

$$-\langle \underline{q}_i \cdot \underline{F}_i \rangle = 3k_B T, \quad i = 1, \ldots, N, \tag{8.19}$$

where $\underline{q}_i \cdot \underline{F}_i$ is the inner product of the two vectors.

In Cartesian coordinates, Eq. 8.19 reads

$$-\langle \underline{r}_i \cdot \underline{F}_i \rangle = 3k_B T. \tag{8.20}$$

Generalizing for N particles, we can write

$$\frac{1}{3} \left\langle \sum_{i=1}^{N} \underline{r}_i \cdot \underline{F}_i \right\rangle = -N k_B T. \tag{8.21}$$

The quantity on the left-hand side of Eq. 8.21 is called the total virial of the system. The total virial can be used to determine the equation of state for non-ideal gases. In what follows we show how.

Let us consider the nature of the force $\underline{F}_i$ exerted on particle i. There are two sources:

1. The force exerted on any particle i by the walls of the system, denoted by $\underline{F}_i^w$.
2. The force exerted on any particle i by the rest $(N - 1)$ of the particles in the system, denoted by $\underline{F}_i^s$.

The total virial can now be written as the sum of two terms, that is

$$\frac{1}{3} \left\langle \sum_{i=1}^{N} \underline{r}_i \cdot \underline{F}_i^w \right\rangle + \frac{1}{3} \left\langle \sum_{i=1}^{N} \underline{r}_i \cdot \underline{F}_i^s \right\rangle = -N k_B T. \tag{8.22}$$

The first term is called the wall virial (Fig. 8.1) and the second term is called the internal virial.

Let us first focus on the wall virial and consider only the x dimension component. If the volume V is a parallelepiped box of size

$L_x \times L_y \times L_z$, and assuming no long-range interactions between the wall and the particles, the wall force on any particle will be non-negligible only at two positions, $x_i = 0$ and $x_i = L_x$.

Consequently,

$$\left\langle \sum_{i=1}^{N} x_i F_{xi}^{w} \right\rangle = L_x \left\langle \sum_{i=1}^{N} F_{xi}^{w} \right\rangle_{x=L_x}. \tag{8.23}$$

The term in the bracket is the average force exerted from the wall on the system at position L. According to Newton's third law, this is equal to the negative average force exerted by the system on the wall. This, in turn, is connected to the pressure of the system:

$$\left\langle \sum_{i=1}^{N} F_{xi}^{w} \right\rangle_{x=L_x} = -PL_yL_z. \tag{8.24}$$

Combining Eq. 8.23 and Eq. 8.24 yields

$$\left\langle \sum_{i=1}^{N} x_i F_{xi}^{w} \right\rangle = -PV. \tag{8.25}$$

Figure 8.1 The wall virial is the sum of the forces of bouncing particles on the wall.

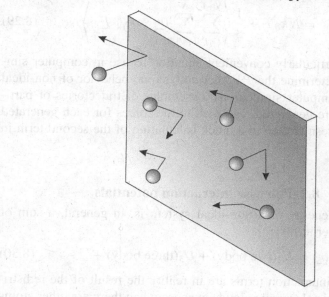

The equation of the total virial can now be written as follows:

$$PV = Nk_BT + \frac{1}{3}\left\langle \sum_{i=1}^{N} \underline{r}_i \cdot \underline{F}_i^s \right\rangle. \tag{8.26}$$

This is a remarkable equation that gives a general expression for the equation of state for any fluid system. The first term on the right-hand side is the well-known ideal gas contribution to the pressure. The second term encompasses the influence of all non-idealities on the pressure and the equation of state of the system. Notably, starting with the ensemble average in the virial theorem, we obtain the thermodynamic behavior of any fluid.

If intermolecular forces can be represented in a pairwise function, then

$$\underline{F}_i^s = \sum_{i \neq j} \underline{F}_{ij}, \tag{8.27}$$

where $\underline{F}_{ij}$ is the force on particle i exerted by particle j.

Using Newton's third law,

$$\underline{F}_{ij} = -\underline{F}_{ji}, \tag{8.28}$$

we can write

$$PV = Nk_BT + \frac{1}{3}\left\langle \sum_{i=1}^{N-1} \sum_{j=i+1}^{N} (\underline{r}_i - \underline{r}_j) \cdot \underline{F}_{ij} \right\rangle. \tag{8.29}$$

This is a particularly convenient equation to use in computer simulations and determine the PV thermodynamic behavior of non-ideal systems. In computer simulations, ensembles or trajectories of particle positions are generated. Calculating the forces for each generated configuration results then in a quick calculation of the second term in Eq. 8.29.

8.2 Pairwise interaction potentials

The potential energy of a non-ideal system is, in general, a sum of interaction order terms

$$U(\underline{r}) = U_2(\text{two body}) + U_3(\text{three body}) + \ldots \tag{8.30}$$

Dispersion interaction terms are in reality the result of the redistribution of atomic electronic clouds in space when there are other atoms close enough. The two-body, or pairwise, interactions dominate the potential energy, except in very low temperature and very high density cases. In these cases the presence of a third particle alters the interaction of two other particles, distorting the electronic clouds and wave functions of these atoms. Discussing three-body or higher order interaction terms is beyond the scope of this text, since the hypothesis of simple pairwise additivity in the potential energy is valid for fluid systems at high temperatures.

We can write, using any one of the equivalent summation formats,

$$U(\underline{r}) = U_2(\text{two body})$$

$$= \sum_{i>j} u(r_{ij})$$

$$= \frac{1}{2} \sum_{i=1}^{N} \sum_{j=1}^{N} u(r_{ij})$$

$$= \sum_{i=1}^{N-1} \sum_{j=i+1}^{N} u(r_{ij}), \qquad (8.31)$$

where r_{ij} is the distance between particles and $u(r_{ij})$ is the interaction potential between them.

Generally, the interaction potential as a function of the distance between two atoms has the form depicted by the dashed-dotted line in Fig. 8.2. Any two non-charged atoms will have zero interaction at sufficiently long distance. As two atoms come close, they always exert

Figure 8.2 Lennard-Jones potential.

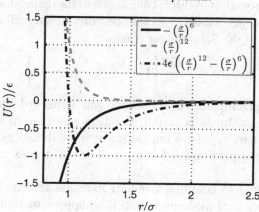

an attractive force on one another. This attractive force, called London dispersion force, or van der Waals attraction, is the result of the momentary polarization of particles. More specifically, the electron cloud of an atom is distorted under the influence of a second atom nearby. This distortion results in the formation of a non-zero dipole on the atom, assuming that the position of the positive nucleus is not affected. There is actually a dipole forming on the second atom as well, because of the first atom. Consequently, the attractive force between any two atoms is the result of inducible dipole–dipole interactions.

At shorter distances, the electron clouds of the two atoms start physically overlapping and interacting with strong Coulombic interactions. A strong repulsive force is then exerted at closer distances.

Two parameters, ε and σ, are typically used to describe mathematically interatomic potentials. Parameter σ is representative of the size of the atoms and is usually the distance where the interaction potential is zero, $u(\sigma) = 0$. Parameter ε is the maximum depth interaction energy, i.e., $ur_{ij} = -\varepsilon$ when $\partial u(r_{ij})/\partial r_{ij} = 0$.

8.2.1 Lennard-Jones potential

Perhaps the most widely used realistic potential of interaction between atoms is the 12/6 Lennard-Jones potential,

$$u(r) = 4\varepsilon \left(\left(\frac{\sigma}{r} \right)^{12} - \left(\frac{\sigma}{r} \right)^{6} \right), \tag{8.32}$$

where we use $r = r_{ij}$ for brevity. The force between any two atoms is then $f = -\partial u(r)/\partial r$.

The sixth power term can be determined theoretically with the help of Schrödinger's equation. A direct calculation of the dispersion interaction between two atoms, modeled as induced dipole interactions, yields a series solution of the Schrödinger equation

$$u_{\text{disp}}(r) = -\frac{C_0}{r^6} \left\{ 1 + \frac{C_1}{r^2} + \frac{C_2}{r^4} + \ldots \right\}. \tag{8.33}$$

For high enough temperatures the second and higher order terms vanish.

In the Lennard-Jones potential, the 12th power term is empirical, modeling the sharp repulsive interaction at small distances.

Lennard-Jones parameters for a range of substances are shown in Table 8.1.

It is interesting to note that Lennard-Jones parameters are assigned even to non-spherical molecules. This is an approximation that is valid for small molecules like oxygen and nitrogen at high temperatures and

Table 8.1 *Lennard-Jones parameters.*

Substance	$\sigma(\text{Å})$	$\varepsilon(k_B)$
H_2	2.915	38.0
He	2.576	10.2
Ne	2.789	35.7
Ar	3.418	124
Kr	3.498	225
Xe	4.055	229
N_2	3.681	91.5
O_2	3.433	113
CO	3.590	110
CO_2	3.996	190
Cl_2	4.115	357
CH_4	3.822	137
C_2H_6	4.418	230
$n - C_9H_2O$	8.448	240

low densities. For larger molecules like $n - C_9H_2O$ this approximation will be of limited accuracy, especially at higher densities.

Although self-evident, it is also worth mentioning that another physical interpretation of the interaction between two particles at distance r is the following: the potential energy $u(r)$ is the reversible work required to move the two particles from distance r to an infinite distance with no interactions. We will see in later chapters that this is actually equal to the free energy of the system of the two particles. For example, parameter ε for argon has a value of $124\,k_B$, which is the reversible work to pull apart two argon atoms originally at distance r where $u(r) = -\varepsilon$.

The Lennard-Jones potential can also be used for interactions between dissimilar atoms. The Berthelot mixing rules can then be used to determine the effective σ and ε parameters. The effective size parameter in the pairwise interaction potential is determined as the arithmetic average of the sizes of the two atoms:

$$\sigma = \frac{\sigma_1 + \sigma_2}{2}, \tag{8.34}$$

where σ_i is the size of atom type i.

The effective potential well depth parameter is determined as the geometric average of the two atoms potential well depth parameters:

$$\varepsilon = \sqrt{\varepsilon_1 \varepsilon_2}, \tag{8.35}$$

where ε_i is the interaction energy parameter of atom type i.

Certainly, there are other potential models that capture the interaction energy between atoms. An example is the Hartree–Fock potential

$$u(r) = A \exp\left(-\frac{C_0 r}{\sigma}\right) + F\left(\frac{C_1}{r^6} + \frac{C_2}{r^8} + \frac{C_3}{r^{10}}\right), \qquad (8.36)$$

where F is a function of the distance

$$F = \begin{cases} \exp\left(-C\left(\frac{r-\sigma}{r}\right)^2\right), & \text{if } r \le C \\ 1. & \text{if } r > C \end{cases} \qquad (8.37)$$

In principle, the Hartree–Fock potential can capture interaction energies very accurately, but clearly this comes at the expense of having six parameters to determine. It also becomes unwieldy to use in computer simulations.

8.2.2 *Electrostatic interactions*

Atoms and molecules may be charged (e.g., chloride and sodium ions), have permanent dipoles (e.g., water), or higher order multipoles (e.g., carbon dioxide is quadrupolar).

Interactions between two charged particles i and j are described by Coulomb's law

$$u_{\text{ele}}(r_{ij}) = \frac{q_i q_j}{4\pi\varepsilon_o \varepsilon r_{ij}}, \qquad (8.38)$$

where q_i and q_j are the charges of particles i and j, respectively, ε_o is the dielectric constant of vacuum, ε is the dielectric constant of any medium between the two charges, and r_{ij} is the distance between them.

Permanent (not induced) electric dipoles in molecules are the result of differential electronegativity, that is propensity to attract electron clouds. For example, in water the oxygen atom has the tendency to attract the electrons of the two hydrogens close to its own nucleus. This results in an effective negative partial charge on the oxygen and two effective positive partial charges (each half the size of the oxygen charge) on the hydrogens. Consequently, a dipole vector forms on every water molecule. The magnitude of the water dipole is 6.2×10^{-30} C m.

Generally, the dipole moment of molecule i is given by

$$\underline{\mu}_i = \sum_k q_k \underline{r}_k, \qquad (8.39)$$

where q_k are the partial charges of the molecule and r_k are the charge positions.

The interaction between two dipoles depends on their separation r and their orientation and is given by

$$u_{\text{dip}} = \frac{1}{4\pi\varepsilon r^3} \left(\underline{\mu}_1 \cdot \underline{\mu}_2 - 3(\underline{\mu}_1 \cdot \underline{r})(\underline{\mu}_2 \cdot \underline{r}) \right). \tag{8.40}$$

This interaction energy applies also to intramolecular interactions between bonds with permanent dipoles.

8.2.3 *Total intermolecular potential energy*

The intermolecular potential energy of a system is the sum of all dispersion and electrostatic particle–particle interactions. In the potential energy there will also be intramolecular terms, such as bonds, bond angles, and torsional interactions, discussed in Chapter 14. In what follows we focus on non-charged monoatomic systems. Consequently, the only term in the potential energy is the dispersion forces term. The results can be readily extended to molecular systems.

8.3 Virial equation of state

In the virial equation of state, the pressure is written as a power series in density, $\rho = N/V$,

$$\frac{P}{k_B T} = \rho + B_2(T)\rho^2 + B_3(T)\rho^3 + \dots, \tag{8.41}$$

where B_2 is the second virial coefficient, associated with pairwise interactions, B_3 is the third virial, associated with three-body interactions, and so on.

It can be shown (this is beyond of the scope of this book; see Further reading texts at the end of this chapter) that the second virial coefficient can be written as a function of the pairwise interaction potential $u(r)$, as follows:

$$B_2(T) = -2\pi \int_0^\infty [\exp(-\beta u(r)) - 1]r^2 dr. \tag{8.42}$$

The virial equation of state for pairwise additive systems is then

$$\frac{P}{k_B T} = \rho - 2\pi\rho^2 \int_0^\infty [\exp(-\beta u(r)) - 1]r^2 dr. \tag{8.43}$$

For Lennard-Jones fluids the second virial can be calculated to be

$$B_2 = -\frac{2\pi\sigma^3}{3} \sum_{n=0}^{\infty} \frac{2^{\frac{2n+1}{2}}}{4n!} \left(\frac{\varepsilon}{k_BT}\right)^{\frac{2n+1}{4}} \Gamma\left(\frac{2n-1}{4}\right), \qquad (8.44)$$

where the Γ function is defined as

$$\Gamma(n+1) = \int_0^{\infty} y^n e^{-y} dy. \qquad (8.45)$$

Naturally, the equation of state of a Lennard-Jones fluid cannot be determined analytically. An analytical expression derivation is feasible for particles interacting with the Sutherland potential, discussed in the next section.

8.4 van der Waals equation of state

Let us consider a non-ideal gas in the NVT ensemble. Assume that the gas particles interact with the Sutherland potential (Fig. 8.3), expressed as:

$$U(r) = \begin{cases} +\infty, & \text{if } r < \sigma \\ -\varepsilon\left(\dfrac{\sigma}{r}\right)^6. & \text{if } r \geq \sigma \end{cases} \qquad (8.46)$$

The second virial is

$$B_2(T) = B(T)$$
$$= -2\pi \int_0^{\sigma} (-1)r^2 dr - 2\pi \int_0^{\infty} r^2 \left[\exp\left(\beta\varepsilon\left(\frac{\sigma}{r}\right)^6\right) - 1\right] dr. \qquad (8.47)$$

Figure 8.3 Sutherland potential.

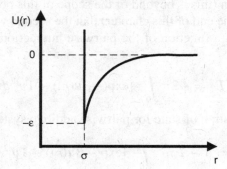

A reasonable assumption for high-temperature systems is that $\varepsilon/k_B T \ll 1$. Then,

$$\exp\left(\beta\varepsilon\left(\frac{\sigma}{r}\right)^6\right) \approx 1 + \beta\varepsilon\left(\frac{\sigma}{r}\right)^6 + \cdots \qquad (8.48)$$

Consequently,

$$B(T) = \frac{2\pi}{3}\sigma^3 - 2\pi\beta\varepsilon\int_\sigma^\infty r^2\left(\frac{\sigma}{r}\right)^6 dr. \qquad (8.49)$$

Then,

$$B(T) = \frac{2\pi}{3}\sigma^3(1 - \beta\varepsilon). \qquad (8.50)$$

The virial equation of state for the pressure yields

$$P = \frac{Nk_B T}{V} + \frac{N^2 k_B T}{V^2}\frac{2\pi}{3}\sigma^3(1 - \beta\varepsilon) \qquad (8.51)$$

or

$$P = \frac{Nk_B T}{V}\left(1 + \frac{2\pi\sigma^3}{3}\rho\left(1 - \frac{\varepsilon}{k_B T}\right)\right), \qquad (8.52)$$

and rearranging

$$\left(P + \frac{2\pi\sigma^3\varepsilon}{3}\rho^2\right) = k_B T\rho\left(1 + \frac{2\pi\sigma^3}{3}\rho\right). \qquad (8.53)$$

Assuming that $N4\pi\sigma^3/3 \ll V$, a reasonable assumption for low-density systems, yields

$$\left(P + \frac{2\pi\sigma^3\varepsilon}{3}\rho^2\right) = k_B T\rho\left(1 - \frac{2\pi\sigma^3}{3}\rho\right)^{-1}. \qquad (8.54)$$

Defining a molecular volume $v = 1/\rho = V/N$ we arrive at the celebrated van der Waals equation of state (Fig. 8.4)

$$\left(P + \frac{a}{v^2}\rho^2\right)(v - b) = k_B T, \qquad (8.55)$$

where

$$\begin{cases} a = \dfrac{2\pi}{3}\sigma^3\varepsilon, \\[2mm] b = \dfrac{2\pi}{3}\sigma^3. \end{cases} \qquad (8.56)$$

van der Waals reached Eq. 8.55 in 1873 with a remarkable set of experiments. We have derived it using statistical thermodynamic arguments, thus providing a microscopic calculation of the vdW parameters.

What is notable regarding the van der Waals volumetric equation of state is that it predicts vapor–liquid phase equilibria for temperatures below the critical temperature. In Fig. 8.4, the pressure–volume behavior is shown at different temperatures as calculated using the van der Waals equation of state. For temperatures higher than the critical temperature, the pressure decreases monotonically with increasing volume. This is the behavior predicted by the ideal gas law. For temperatures lower than the critical point, the pressure drops with increasing volume until it reaches a minimum. The PV isotherm slope then changes with the pressure, increasing until it reaches a maximum. The slope changes yet again and the pressure continues then to drop as the volume increases.

Therefore, below the critical temperature, the van der Waals equation of state predicts three distinct systems with different molar volumes at the same pressure and temperature. It turns out that only the two molar volumes at the ends are thermodynamically stable (an elegant proof of the instability of the middle molar volume is given in Sandler's handbook, see Further reading). The lower molar volume represents a liquid state, whereas the higher molar volume represents a vapor gas state. These two states are in equilibrium.

We discuss the properties of the van der Waals equation in more detail in Chapter 10.

Figure 8.4 Isotherms for van der Waals equation of state. $P_r = P/P_C$ and $V_r = V/V_C$, where P_C and V_C are the critical pressure and volume, respectively.

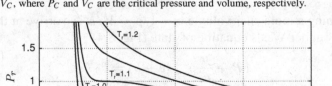

8.5 Further reading

1. S. I. Sandler, *Chemical and Engineering Thermodynamics*, (New York: Wiley, 1999).
2. J. M. Smith, H. C. Van Ness, and M. M. Abbott, *Introduction to Chemical Engineering Thermodynamics*, (New York: McGraw Hill, 2005).
3. H. Metiu, *Physical Chemistry, Thermodynamics*, (New York: Taylor and Francis, 2006).
4. D. A. McQuarrie, *Statistical Mechanics*, (Sausalito, CA: University Science Books, 2000).
5. T. L. Hill, *Statistical Thermodynamics*, (Reading, MA: Addison-Wesley, 1960).

8.6 Exercises

1. Show that for particles interacting with a pairwise potential $u(r)$, the distribution of particles around any selected particle is given by the Boltzmann formula

$$\rho \exp\left(-\frac{1}{k_B T}\int_r^\infty u(r)dr\right),$$

where $\rho = N/V$ is the number density.

2. Find the second virial coefficient for the Berthelot equation of state

$$\left(P + \frac{N^2 A}{V^2 T}\right)(V - NB) = Nk_B T,$$

where A and B are constants.

3. Show that

$$B_2 = -\frac{1}{6kT}\int_0^\infty r\frac{du(r)}{dr}e^{-u(r)/kT}4\pi r^2 dr$$

is equivalent to

$$B_2 = -\frac{1}{2}\int_0^\infty \left(e^{-u(r)/kT} - 1\right)4\pi r^2 dr.$$

State the condition on $u(r)$ that is necessary.

4. The second virial coefficients of perfluorohexane, $n\text{-}C_6F_{14}$, are shown in the following table

Temperature, (K)	$B_2(T)$, (lt/mol)
395.6	−1.051
415.5	−0.920
432.7	−0.818
451.5	−0.725

Determine the Lennard-Jones parameters for perfluorohexane and the second virial coefficient you would calculate with these parameters. Compare with experiment.

5. An approximate partition function for a dense gas has the following form:

$$Q(N, V, T) = \frac{1}{N!} \left(\frac{2\pi m k_B T}{h^2} \right)^{3N/2} (V - bN)^N \exp\left(\frac{aN^2}{V k_B T} \right),$$

where a and b are constants that are given in terms of molecular parameters. Calculate the equation of state from this partition function. What equation of state is it? Calculate the thermodynamic energy and the heat capacity and compare them to the ideal gas ones.

6. Use the Lennard-Jones parameters of O_2 in Table 8.1 to predict the constants defining its vapor–liquid critical point. Find these parameters in the NIST Chemistry WebBook (*http://webbook.nist.gov/chemistry/fluid/*) and compare.

Liquids and crystals

9.1 Liquids

The determination of thermodynamic properties of systems with interacting particles is challenging because of the intractability of the configurational integral calculation,

$$Z(N, V, T) = \int d\underline{q} \, \exp(-\beta U(\underline{q})). \tag{9.1}$$

To determine $Z(N, V, T)$, all microscopic configurations must be accounted for, and the energy determined for each. For systems of low density, because of the relatively weak interactions, we were able to introduce reasonable assumptions and to avoid the direct computation of $Z(N, V, T)$. We will also see that for solid systems, because of the highly ordered state of the particles, reasonable assumptions can again be made that simplify calculation of the configurational integral and the partition function.

In contrast, for systems in the liquid state, there are no such simple, valid assumptions to be made. Consequently, the great disorder and the strong interactions between particles render calculation of the configurational integral, in Eq. 9.1, intractable. Different approaches are then required to tackle the theoretical derivation of thermodynamics properties of liquids.

John G. Kirkwood (1907–1959) was the first scientist to develop a complete statistical mechanical theory for liquids. Kirkwood used the concept of the molecular distribution function to develop his liquid state theory during the 1930s and early 1940s. In the next section, we introduce the required background for determining thermodynamic properties of liquids.

9.2 Molecular distributions

Consider the NVT ensemble of a monoatomic system in the liquid phase. The kinetic energy component of the Hamiltonian, which depends on the momenta of the particles, can still be integrated in the partition function. This yields for the partition function

$$Q(N, V, T) = \frac{1}{N!} \left(\frac{2\pi m k_B T}{h^2} \right)^{3N/2} Z(N, V, T). \qquad (9.2)$$

Thus, we can now concentrate on the configurational space, i.e., the $3N$-dimensional subspace of phase space Γ for the positions of the atoms.

Each point in configurational space describes a particular microscopic state of atom positions $(\underline{r}_1, \underline{r}_2, \ldots, \underline{r}_N)$.

The probability density in configurational space is

$$\rho(\underline{r}_1, \underline{r}_2, \ldots, \underline{r}_N) = \frac{\exp[-\beta U(\underline{r}_1, \underline{r}_2, \ldots, \underline{r}_N)]}{\int \exp[-\beta U(\underline{r}_1, \underline{r}_2, \ldots, \underline{r}_N)] d\underline{r}_1 d\underline{r}_2 \ldots d\underline{r}_N} \qquad (9.3)$$

or

$$\rho(\underline{r}) = \frac{\exp[-\beta U(\underline{r})]}{Z}. \qquad (9.4)$$

Equation 9.4 yields the probability density of any particular configuration of atoms in space V. Let us momentarily assume the particles are distinguishable. The probability of finding

$$\begin{cases} \text{particle 1 at position } \underline{r}_1 \text{ to } \underline{r}_1 + d\underline{r}_1, \\ \text{particle 2 at position } \underline{r}_2 \text{ to } \underline{r}_2 + d\underline{r}_2, \\ \ldots, \text{ and} \\ \text{particle } N \text{ at position } \underline{r}_N \text{ to } \underline{r}_N + d\underline{r}_N, \end{cases}$$

is then

$$P(\underline{r}_1, \underline{r}_2, \ldots, \underline{r}_N) = \rho(\underline{r}_1, \underline{r}_2, \ldots, \underline{r}_N) d\underline{r}_1 d\underline{r}_2 \ldots d\underline{r}_N. \qquad (9.5)$$

We define the probability density of finding particle 1 at $\underline{r}_1$ to $\underline{r}_1 + d\underline{r}_1$ irrespective of the rest as

$$\rho^{(1)}(\underline{r}_1) = \int \rho(\underline{r}_1, \underline{r}_2, \ldots, \underline{r}_N) d\underline{r}_2 \ldots d\underline{r}_N \qquad (9.6)$$

or

$$\rho^{(1)}(\underline{r}_1) = \frac{1}{Z} \int \exp[-\beta U(\underline{r})] d\underline{r}_2 \ldots d\underline{r}_N. \qquad (9.7)$$

Note that the integration in Eqs. 9.6 and 9.7 runs over all particles from 2 to N.

The corresponding probability is

$$P^{(1)}(\underline{r}_1) = \rho^{(1)}(\underline{r}_1)d\underline{r}_1. \tag{9.8}$$

More generally, the probability density of finding

$$\begin{cases} \text{particle 1 at position } \underline{r}_1 \text{ to } \underline{r}_1 + d\underline{r}_1, \\ \text{particle 2 at position } \underline{r}_2 \text{ to } \underline{r}_2 + d\underline{r}_2, \\ \ldots, \text{ and} \\ \text{particle } n \text{ at position } \underline{r}_n \text{ to } \underline{r}_n + d\underline{r}_n, \end{cases}$$

where $n < N$, irrespective of the other $n + 1, \ldots, N$ particles, is

$$\rho^{(1,\ldots,n)}(\underline{r}_1, \underline{r}_2, \ldots, \underline{r}_n) = \int \rho(\underline{r}_1, \underline{r}_2, \ldots, \underline{r}_N)d\underline{r}_{n+1} \ldots d\underline{r}_N. \tag{9.9}$$

The corresponding probability is

$$P^{(1,\ldots,n)}(\underline{r}_1, \underline{r}_2, \ldots, \underline{r}_n) = \rho^{(1,\ldots,n)}(\underline{r}_1, \underline{r}_2, \ldots, \underline{r}_n)d\underline{r}_1 d\underline{r}_2 \ldots d\underline{r}_n. \tag{9.10}$$

Next, we define the probability density of finding any one particle at $\underline{r}_1$ to $\underline{r}_1 + d\underline{r}_1$, irrespective of the rest as

$$\rho_N^{(1)}(\underline{r}_1) = N\rho^{(1)}(\underline{r}_1), \tag{9.11}$$

or, using the configurational integral

$$\rho_N^{(1)}(\underline{r}_1) = \frac{N}{Z} \int \exp[-\beta U(\underline{r})]d\underline{r}_2 \ldots d\underline{r}_N. \tag{9.12}$$

We call $\rho_N^{(1)}(\underline{r}_1)$ the one-particle density.
Generally, the probability density of finding

$$\begin{cases} \text{any particle at position } \underline{r}_1 \text{ to } \underline{r}_1 + d\underline{r}_1, \\ \text{any other particle at position } \underline{r}_2 \text{ to } \underline{r}_2 + d\underline{r}_2, \\ \ldots, \text{ and} \\ \text{any particle at position } \underline{r}_n \text{ to } \underline{r}_n + d\underline{r}_n, \end{cases}$$

irrespective of the other $n + 1, \ldots, N$ particles, is

$$\rho_N^{(n)}(\underline{r}_1, \underline{r}_2, \ldots, \underline{r}_n) = N(N - 1) \ldots (N - n + 1)\rho^{(1\ldots n)}(\underline{r}_1, \underline{r}_2, \ldots, \underline{r}_n) \tag{9.13}$$

or

$$\rho_N^{(n)}(\underline{r}_1, \underline{r}_2, \ldots, \underline{r}_n) = \frac{N!}{(N - n)!}\rho^{(1\ldots n)}(\underline{r}_1, \underline{r}_2, \ldots, \underline{r}_n). \tag{9.14}$$

The density $\rho_N^{(n)}$ is called the *n*-particle density.

The corresponding probability is

$$P_N^{(n)}(\underline{r}_1, \underline{r}_2, \ldots, \underline{r}_n) = \rho_N^{(n)}(\underline{r}_1, \underline{r}_2, \ldots, \underline{r}_n) d\underline{r}_1 d\underline{r}_2 \ldots d\underline{r}_n \quad (9.15)$$

or

$$P_N^{(n)}(\underline{r}_1, \underline{r}_2, \ldots, \underline{r}_n) = \frac{N!}{(N-n)!} P^{(1\ldots n)}(\underline{r}_1, \underline{r}_2, \ldots, \underline{r}_n). \quad (9.16)$$

Let us now define the n-particle distribution function as

$$g_N^{(n)}(\underline{r}_1, \underline{r}_2, \ldots, \underline{r}_n) = \frac{\rho_N^{(n)}(\underline{r}_1, \underline{r}_2, \ldots, \underline{r}_n)}{\rho^n}, \quad (9.17)$$

where $\rho = N/V$ is the particle number density of the system.

Of interest in the derivation of thermodynamic properties is the pairwise distribution function

$$g_N^{(2)}(\underline{r}_1, \underline{r}_2) = \frac{\rho_N^{(2)}(\underline{r}_1, \underline{r}_2)}{\rho^2}. \quad (9.18)$$

In Eq. 9.18, $\rho_N^{(2)}(\underline{r}_1, \underline{r}_2)$ is the two-particle probability density of finding

$$\begin{cases} \text{any particle at position } \underline{r}_1 \text{ to } \underline{r}_1 + d\underline{r}_1, \\ \text{and any other particle at position } \underline{r}_2 \text{ to } \underline{r}_2 + d\underline{r}_2. \end{cases}$$

This can be written as

$$\rho_N^{(2)}(\underline{r}_1, \underline{r}_2) = N(N-1)\rho^{(1,2)}(\underline{r}_1, \underline{r}_2) \quad (9.19)$$

or

$$\rho_N^{(2)}(\underline{r}_1, \underline{r}_2) = N(N-1)\frac{\int \exp[-\beta U(\underline{r})] d\underline{r}_3 \ldots d\underline{r}_N}{Z}. \quad (9.20)$$

The pair distribution function of a monoatomic liquid is then defined as follows,

$$g_N^{(n)}(\underline{r}_1, \underline{r}_2, \ldots, \underline{r}_n) = \frac{N(N-1)}{\rho^2} \frac{\int \exp[-\beta U(\underline{r})] d\underline{r}_3 \ldots d\underline{r}_N}{Z}. \quad (9.21)$$

9.3 Physical interpretation of pair distribution functions

Kirkwood brilliantly realized that liquids are not completely random distributions of atoms in volume V (Fig. 9.1). Instead liquids possess some residual local structure because of pairwise interactions. He also understood that this local structure can be quantified with the help of the pair distribution function. Let us determine how.

In the previous section, we found that the joint probability of finding

$$\begin{cases} \text{particle 1 at position } \underline{r}_1 \text{ to } \underline{r}_1 + d\underline{r}_1, \\ \text{and particle 2 at position } \underline{r}_2 \text{ to } \underline{r}_2 + d\underline{r}_2 \end{cases}$$

is

$$\begin{aligned} P^{(2)}(\underline{r}_1, \underline{r}_2) &= \rho^{(2)}(\underline{r}_1, \underline{r}_2) d\underline{r}_1 d\underline{r}_2 \\ &= \frac{1}{N(N-1)} \rho_N^{(2)}(\underline{r}_1, \underline{r}_2) d\underline{r}_1 d\underline{r}_2. \end{aligned} \tag{9.22}$$

This can be expressed in terms of the pairwise distribution function as follows:

$$P^{(2)}(\underline{r}_1, \underline{r}_2) = \frac{\rho^2 g_N^{(2)}(\underline{r}_1, \underline{r}_2)}{N(N-1)} d\underline{r}_1 d\underline{r}_2. \tag{9.23}$$

Consequently, the conditional probability $P^{(2)}(\underline{r}_2 | \underline{r}_1)$ of finding

$$\begin{cases} \text{particle 2 at position } \underline{r}_2 \text{ to } \underline{r}_2 + d\underline{r}_2, \\ \text{provided that particle 1 is at position } \underline{r}_1 \text{ to } \underline{r}_1 + d\underline{r}_1 \end{cases}$$

can be calculated by dividing the joint probability in Eq. 9.22 with the probability of finding particle 1 at position $\underline{r}_1$ to $\underline{r}_1 + d\underline{r}_1$.

Figure 9.1 Liquid structure.

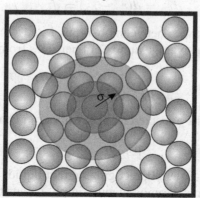

The latter probability is simply

$$P(\underline{r}_1) = \frac{d\underline{r}_1}{\int d\underline{r}_1}. \tag{9.24}$$

Therefore

$$
\begin{aligned}
P^{(2)}(\underline{r}_2|\underline{r}_1) &= \frac{P^{(2)}(\underline{r}_1, \underline{r}_2)}{P(\underline{r}_1)} \\
&= \frac{\rho^2 g_N^{(2)}(\underline{r}_1, \underline{r}_2) d\underline{r}_1 d\underline{r}_2}{\dfrac{N(N-1) d\underline{r}_1}{\int d\underline{r}_1}}.
\end{aligned} \tag{9.25}
$$

This means that the average number of particles at position $\underline{r}_2$ to $\underline{r}_2 + d\underline{r}_2$, provided that there is a particle at position $\underline{r}_1$ to $\underline{r}_1 + d\underline{r}_1$, is the conditional probability $P^{(2)}(\underline{r}_2|\underline{r}_1)$ for any particle, which is

$$(N-1) \frac{\rho^2 g_N^{(2)}(\underline{r}_1, \underline{r}_2) d\underline{r}_1 d\underline{r}_2}{\dfrac{N(N-1) d\underline{r}_1}{\int d\underline{r}_1}}. \tag{9.26}$$

Note that $\rho = N/V$, and that

$$\int d\underline{r}_1 = V. \tag{9.27}$$

Finally, note that for a given $\underline{r}_1$

$$d\underline{r}_2 = d(\underline{r}_2 - \underline{r}_1) = d\underline{r}_{12}, \tag{9.28}$$

where $\underline{r}_{12}$ is the distance between positions 1 and 2.

Consequently, the average number of particles from $\underline{r}_{12}$ to $\underline{r}_{12} + d\underline{r}_{12}$ is

$$\rho g_N^{(2)}(\underline{r}_1, \underline{r}_2) d\underline{r}_{12}. \tag{9.29}$$

In other words, the expected local density around a particle is

$$\rho g_N^{(2)}(\underline{r}_1, \underline{r}_2). \tag{9.30}$$

We can simplify Eq. 9.30, recognizing that liquids are homogeneous and isotropic systems. Homogeneous simply means that their structure depends only on distance. We can then write for the pair distribution function

$$g_N^{(2)}(\underline{r}_1, \underline{r}_2) = g_N^{(2)}(\underline{r}_{12}).$$ (9.31)

Isotropic means that the direction is not important. We can then write for the pair distribution function

$$g_N^{(2)}(\underline{r}_{12}) = g_N^{(2)}(r_{12}).$$ (9.32)

Therefore we can simply and concisely use $g(r)$ to denote the pair distribution function of liquids.

If we denote the local density with $\rho(r)$, then Eq. 9.30 yields

$$\rho(r) = \rho g(r).$$ (9.33)

This is a particularly important result that reveals the physical significance of pair distribution functions (pdf). The pdf of a liquid describes the deviation from a random structure in a small neighborhood around any particle. If $\rho(r) = \rho$, or $g(r) = 1$, even for very small distances of the order of a particle diameter, then around any particle there would be N/V particles even as $V \to 0$. This is physically untenable since the particle itself has a finite size, or a finite excluded volume, where other particles cannot be found lest the potential energy become infinite. Consequently, it is expected that $\rho(r)$ is zero at $r = 0$ and remains insignificant until distances about equal to the particle diameter.

Let us denote the particle diameter by σ. It turns out that in liquids there is a shell of particles at a distance of around σ away from any particle (Fig. 9.1), so that the local density $\rho(\sigma)$ is higher than ρ. Naturally, because of the excluded volume of the particles in the first shell, the local density drops between σ and around 2σ. A second shell of particles is then observed at 2σ that again increases the local density above ρ. For this second shell, the increase is not as significant as for the first shell. Then the density drops again and rises for a third shell.

Figure 9.2 illustrates a typical liquid pdf which captures the short-range order of liquids. It is clear that this characteristic short-range order in liquids is the effect of the impenetrability of close-packed particles. At high densities, when the position of a particle is given, other particles will inevitably cluster around the first in structured layers. After two or three layers, this inherent structuring weakens, and the liquid attains its intuitive disorderly state. Morrell and Hilderbrand illustrated this nicely in a 1936 paper (see Further reading). They used dyed gelatin spheres and placed them in a close-packed arrangement. Photographs of multiple experiments with different random arrangements clearly revealed a short-range order, duplicating the salient characteristics of the molecular distributions of liquids.

How then can this information about the local structure of fluids be used to determine thermodynamic properties? John Kirkwood was able to find an ingenious theoretical path from the pairwise potential of interaction between two particles, $u(r)$, and the pair distribution function to the internal energy and the pressure of liquids. In Section 9.4, we briefly present this theoretical path. Note that it is beyond the scope of this text to discuss the details of Kirkwood's efforts to determine $g(r)$ with purely theoretical arguments. The interested reader should consult Kirkwood's papers for a detailed exposition of the theory (Further reading). We should also note that pair distribution functions of materials can be readily obtained with diffraction measurements and, as discussed later in the book, using computer simulations.

Of importance at this point is to demonstrate how the thermodynamic behavior of a liquid can be derived if the pair distribution function is known. We do this in the following section.

9.4 Thermodynamic properties from pair distribution functions

For any non-ideal system, the excess internal energy can be defined as

$$\langle E^{\text{excess}} \rangle = \langle E(N, V, T) \rangle - \langle E^{IG}(N, V, T) \rangle, \qquad (9.34)$$

where $E^{IG}(N, V, T)$ is the ideal gas energy.

The excess energy encompasses all the interactions between particles

$$\langle E^{\text{excess}} \rangle = \langle U(\underline{r}_1, \underline{r}_2, \ldots, \underline{r}_N) \rangle, \qquad (9.35)$$

where U is the potential energy.

Figure 9.2 Liquid pair distribution function.

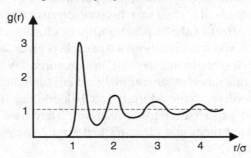

Assuming pairwise interactions, we write

$$\langle U(\underline{r}_1, \underline{r}_2, \ldots, \underline{r}_N) \rangle = \left\langle \sum_{i<j} u(r_{ij}) \right\rangle, \tag{9.36}$$

where $u(r_{ij})$ is the pairwise interaction potential. This typically includes dispersion and electrostatic interactions, as well as internal degree of freedom terms, such as bond and angle potentials.

The pair distribution function gives the probability of finding a second particle in a given volume element at position r from a first particle. Specifically, in volume

$$d\underline{r} = 4\pi r^2 dr, \tag{9.37}$$

the average number of particles found is

$$dN = g(r)\rho 4\pi r^2 dr. \tag{9.38}$$

The contribution of this differential system to the potential energy is then

$$u(r)g(r)\rho 4\pi r^2 dr. \tag{9.39}$$

The total contribution of a particle to the potential energy is the integral of the differential energy in Eq. 9.39 over all distances r, or

$$\int_0^\infty u(r)g(r)\rho 4\pi r^2 dr. \tag{9.40}$$

Consequently, we can determine the total excess energy for all molecules by multiplying Eq. 9.40 with $N/2$, or

$$\langle E^{\text{excess}} \rangle = 2\pi N\rho \int_0^\infty u(r)g(r)r^2 dr. \tag{9.41}$$

We can turn to the pressure and in an analogous manner write

$$PV = (PV)^{IG} + (PV)^{\text{excess}} \tag{9.42}$$

where the term $(PV)^{\text{excess}}$ is associated with interatomic forces.

Using similar arguments as for the calculation of the excess energy, this term can be found to be

$$(PV)^{\text{excess}} = \frac{2\pi N\rho}{3} \int_0^\infty \left(-\frac{\partial u(r)}{\partial r} \right) g(r)r^3 dr. \tag{9.43}$$

Therefore, if the pair distribution function of a non-ideal NVT system is known, then the internal energy and the pressure can be determined as follows:

$$U = \frac{3}{2}Nk_BT + 2\pi N\rho \int_0^\infty u(r)g(r)r^2 dr \qquad (9.44)$$

and

$$PV = Nk_BT - \frac{2\pi N\rho}{3} \int_0^\infty \left(\frac{\partial u(r)}{\partial r}\right) g(r)r^3 dr. \qquad (9.45)$$

These are remarkable relations. With knowledge of only the structure of a liquid, as captured by the pair distribution function, $g(r)$, the macroscopic, thermodynamic behavior of a liquid is completely determined. Starting from $g(r)$, the internal energy and the pressure can be calculated. According to Gibbs' phase rule, for single phase, single-component systems, the values of these two thermodynamic properties completely dictate the thermodynamic state of the material. de Boer and Uhlenbeck provide a critical study of theories of the liquid state (see Further reading). Presenting experimentally determined thermodynamic properties of simple substances over a wide range of conditions, they conclude that there is remarkable agreement between theoretically predicted and experimentally determined properties for simple systems.

9.5 Solids

For solids, statistical mechanics arguments can be used to determine the structure and thermodynamic properties of interest, such as the energy, heat capacity and heat conductivity. In this section, we present simple models to determine the heat capacity of solid crystal structures. For a broader exposition of statistical mechanical theories pertaining to solid materials, the reader is referred to the literature cited at the end of the chapter.

9.5.1 Heat capacity of monoatomic crystals

Consider N atoms of mass m in a solid crystal. We can assume the size of the system to be large enough and the geometry to be such that the surface to volume ratio of the system is vanishingly small. The crystal can be considered as a lattice with regularly spaced atoms (Fig. 9.3). The configuration of the solid is described by the $3N$-dimensional vector of the positions of the atoms

$$X = X(\underline{r}_1, \underline{r}_2, \ldots, \underline{r}_N), \qquad (9.46)$$

where $\underline{r}_i = (x_i, y_i, z_i)$ are the Cartesian coordinates of atom i. Each atom has a conjugate momentum $\underline{p}_i = (px_i, py_i, pz_i)$.

The Hamiltonian of the system is

$$H(\underline{p}, \underline{r}) = K(\underline{p}) + U(\underline{r})$$

$$= \sum_{i=1}^{3N} \frac{p_i^2}{2m} + U(\underline{r}_1, \underline{r}_2, \ldots, \underline{r}_N). \qquad (9.47)$$

At equilibrium each atom i is positioned at lattice site $\underline{ro}_i$. By definition, the potential energy becomes minimum at equilibrium, i.e.,

$$\frac{\partial U(\underline{r}_1, \underline{r}_2, \ldots, \underline{r}_N)}{\partial \underline{r}_i}\bigg|_{\underline{r}_i = \underline{ro}_i} = 0. \qquad (9.48)$$

We can assume that atoms vibrate independently around their equilibrium positions (Fig. 9.4). We then define the displacement $\delta \underline{r}_i$ for each atom i as follows:

$$\delta \underline{r}_i = \underline{r}_i - \underline{ro}_i. \qquad (9.49)$$

Figure 9.3 Crystal structure.

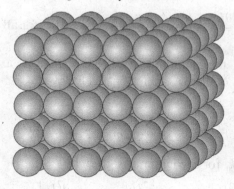

Figure 9.4 Harmonic crystal vibrations.

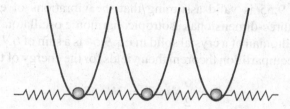

Consequently, the potential energy of the system becomes

$$U(\underline{ro}_1 + \delta\underline{r}_1, \underline{ro}_2 + \delta\underline{r}_2, \ldots, \underline{ro}_N + \delta\underline{r}_N). \tag{9.50}$$

In general, the vibrations can be assumed small for temperatures lower than the melting point of the solid. A Taylor series expansion can then be constructed for the potential energy up to quadratic terms of $\delta\underline{r}_i$ as follows:

$$
\begin{aligned}
U(\underline{r}_1, \underline{r}_2, \ldots, \underline{r}_N) &= U(\underline{ro}_1, \underline{ro}_2, \ldots, \underline{ro}_N) \\
&+ \sum_{i=1}^{N} \left. \frac{\partial U(\underline{r}_1, \underline{r}_2, \ldots, \underline{r}_N)}{\partial \underline{r}_i} \right|_{\underline{r}_i = \underline{ro}_i} \delta\underline{r}_i \\
&+ \frac{1}{2} \sum_{i=1}^{N} \left. \frac{\partial^2 U(\underline{r}_1, \underline{r}_2, \ldots, \underline{r}_N)}{\partial \underline{r}_i^2} \right|_{\underline{r}_i = \underline{ro}_i} \delta\underline{r}_i^2. \tag{9.51}
\end{aligned}
$$

By definition, at equilibrium the potential energy is zero,

$$U(\underline{ro}_1, \underline{ro}_2, \ldots, \underline{ro}_N) = 0. \tag{9.52}$$

So is the first derivative of the potential energy around the lattice site positions,

$$\left. \frac{\partial U(\underline{r}_1, \underline{r}_2, \ldots, \underline{r}_N)}{\partial \underline{r}_i} \right|_{\underline{r}_i = \underline{ro}_i} = 0. \quad \text{for all } i \tag{9.53}$$

Consequently, the Hamiltonian can be written as a function of N momenta $\underline{p}_i$ and N displacements $\delta\underline{r}_i$, as follows:

$$H(\underline{p}, \underline{r}) = \sum_{i=1}^{N} \frac{p_i^2}{2m} + \frac{1}{2} \sum_{i=1}^{N} k_i \delta\underline{r}_i^2. \tag{9.54}$$

The constant k_i for each atom i is given by

$$k_i = \left. \frac{\partial^2 U}{\partial x_i^2} \right|_{x_i = xo_i} = \left. \frac{\partial^2 U}{\partial y_i^2} \right|_{y_i = yo_i} = \left. \frac{\partial^2 U}{\partial z_i^2} \right|_{z_i = zo_i}. \tag{9.55}$$

Equation 9.55 is valid assuming that the vibrations of each atom resemble a three-dimensional, isotropic, harmonic oscillator.

The Hamiltonian of a crystal solid in Eq. 9.54 is a sum of $6N$ quadratic terms. The equipartition theorem then yields for the energy of the system

$$E = \frac{6}{2} N k_B T = 3 N k_B T. \tag{9.56}$$

The constant volume heat capacity is consequently

$$C_V = 3Nk_B. \tag{9.57}$$

This is the celebrated Petit–Dulong law. Pierre Dulong and Alexis Petit proposed this law in 1819, based on a series of careful experiments with numerous different types of solid. The law simply states that the heat capacity is the same for all solids that can be modeled with a crystal lattice. According to Petit and Dulong, the universal value of the molar heat capacity is 24.94 J/mol K. It turns out that this law is valid for high temperatures that are well below the melting point of solids. For example the heat capacity of copper at room temperature is 24.6 J/mol K and for lead 26.5 J/mol K.

9.5.2 The Einstein model of the specific heat of crystals

In 1907, Einstein proposed a simple molecular model for solids that reproduces the Petit–Dulong law. He hypothesized that the vibrations of all N atoms in all three dimensions have the same frequency ν_o. Each atom is then a simple oscillator, which, according to quantum mechanics, has energy levels

$$\varepsilon_n = \left(n + \frac{1}{2} \right) h\nu_o. \quad n = 0, 1, 2 \ldots \tag{9.58}$$

The oscillator partition function is

$$q = \sum_{n=0}^{\infty} e^{-\beta \varepsilon_n}. \tag{9.59}$$

Using Eq. 9.58, the partition function is evaluated as

$$q = \frac{e^{-\beta h\nu_o/2}}{(1 - e^{-\beta h\nu_o})}. \tag{9.60}$$

The crystal lattice partition function is then

$$Q = q^{3N} = \left(\frac{e^{-\beta h\nu_o/2}}{(1 - e^{-\beta h\nu_o})} \right)^{3N}. \tag{9.61}$$

The internal energy of the solid is again

$$U = k_B T^2 \frac{\partial \ln Q}{\partial T}. \tag{9.62}$$

Simple algebraic manipulations yield the following expression for the internal energy,

$$U = U_o + \frac{3Nh v_o e^{-\beta h v_o}}{1 - e^{-\beta h v_o}}. \tag{9.63}$$

In Eq. 9.63, U_o is the zero-point energy of the crystal for $n = 0$, simply given as

$$U_o = \frac{3Nh v_o}{2}. \tag{9.64}$$

The entropy of a perfect crystal can now be calculated as follows:

$$S = \frac{U}{T} + k_B \ln Q \tag{9.65}$$

or

$$S = 3N k_B \left[\frac{\beta h v_o}{e^{\beta h v_o} - 1} - \ln \left(1 - e^{\beta h v_o} \right) \right]. \tag{9.66}$$

According to Eq. 9.66, in the limit of $T \to 0$, the entropy $S \to 0$ as well, which conforms with Planck's statement of the third law of thermodynamics. A discussion of the implications of the third law of thermodynamics, and indeed of whether this law should be regarded as fundamental in the same way as the first two laws of thermodynamics, is beyond the scope of this book.

Returning to the determination of thermodynamic properties of solids, we use Eq. 9.63 to compute the heat capacity of a solid crystal as follows:

$$C_V = \frac{\partial U}{\partial T} = \frac{3Nh v_o e^{\beta h v_o}(h v_o / k_B T^2)}{(e^{\beta h v_o} - 1)^2}. \tag{9.67}$$

Einstein introduced a characteristic temperature parameter

$$\Theta_E = \frac{h v_o}{k_B}. \tag{9.68}$$

A dimensionless heat capacity then takes the form

$$\frac{C_V}{N k_B} = \frac{3 \left(\dfrac{\Theta_E}{T} \right)^2 e^{\Theta_E / T}}{\left(e^{\Theta_E / T} - 1 \right)^2}. \tag{9.69}$$

This is a remarkable relation that resembles a law of corresponding states for the heat capacity of crystalline solids. Different crystals have the same heat capacity at corresponding temperatures.

In Fig. 9.5, the Einstein model heat capacity is shown as a function of the characteristic temperature. Equation 9.69 correctly predicts the heat capacity at the limits of $T \to 0$ and $T \to \infty$.

As the temperature goes to zero, $\Theta_E/T \to \infty$. Then $\left(e^{\Theta_E/T} - 1\right)^2 \to e^{2\Theta_E/T}$. Consequently, the heat capacity is

$$\frac{C_V}{Nk_B} \to 0, \tag{9.70}$$

which correctly captures experimental observations.

As the temperature increases, at the limit $T \to \infty$, $\Theta_E/T \to 0$, $e^{\Theta_E/T} \to 1$, and $\left(e^{\Theta_E/T} - 1\right) \to \Theta_E/T$. Consequently, the heat capacity becomes

$$\frac{C_V}{Nk_B} \to 3, \tag{9.71}$$

recovering the Petit–Dulong law.

Although Einstein's model correctly captures the heat capacity at the limits of very low and very high temperatures, it agrees poorly with the observed heat capacity of solids for a wide range of temperatures. This is because of the assumption that all vibrations in the crystal have the same frequency.

9.5.3 The Debye model of the specific heat of crystals

In 1912, Debye further improved on the model by assuming that particle vibration frequencies are continuously distributed over a range from 0 to a maximum frequency ν_m. If the number of vibrational frequencies

Figure 9.5 The Einstein heat capacity.

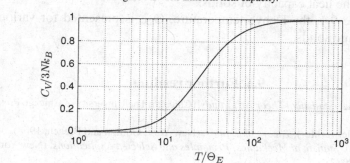

in the range between v and $v + dv$ is $f(v)dv$, the partition function of the crystal can be written as

$$\ln Q = -\frac{U_o}{k_B T} \int_0^{v_m} \ln\left(1 - e^{\beta h v}\right) f(v)dv. \qquad (9.72)$$

Debye used the theory of elastic vibrations in continuous solids to postulate that

$$f(v) = \begin{cases} 9N v_m^{-3} v^2, & \text{if } 0 \leq v \leq v_m \\ 0. & \text{if } v > v_m \end{cases} \qquad (9.73)$$

Consequently, the heat capacity is determined to be

$$\frac{C_V}{k_B T} = 9\left(T/\Theta_D\right)^3 \int_0^{\Theta_D/T} \frac{y^4 e^y}{(e^y - 1)^2} dy \qquad (9.74)$$

where y is the integration variable, and the Debye characteristic temperature is

$$\Theta_D = \frac{h v_m}{k_B}. \qquad (9.75)$$

At high temperatures the heat capacity according to the Debye model is

$$\frac{C_V}{k_B T} = 3. \qquad (9.76)$$

At low temperatures, the heat capacity becomes

$$\frac{C_V}{k_B T} = \frac{12\pi^2}{5} \left(\frac{T}{\Theta_D}\right)^3. \qquad (9.77)$$

This is the Debye T^3 law, which gives an accurate temperature dependence of the heat capacity of solids.

In Table 9.1, the Debye temperature, Θ_D, is presented for various monoatomic solids.

9.6 Further reading

1. I. Z. Fisher, *Statistical Theory of Liquids*, (Chicago: The University of Chicago Press, 1964).
2. B. J. Adler (Editor), *Theory of Liquids*, (New York: Gordon and Breach, 1968).
3. T. L. Hill, *Statistical Mechanics: Principles and Selected Applications*, (New York: Dover, 1987).

Table 9.1 *Debye characteristic temperature.*

Solid	$\Theta_D(°K)$
Ag	215
Al	390
Au	170
Be	1000
C (diamond)	1860
Co	385
Cu	315
Fe	420
Mg	290
Ni	375
Pb	88
Pt	225
Zn	250

4. L. A. Girifalco, *Statistical Mechanics of Solids*, (London: Oxford University Press, 2003).
5. D. A. McQuarrie, *Statistical Mechanics*, (Sausalito, CA: University Science Books, 2000).
6. J. de Boer and G. E. Uhlenbeck (Editors), *Studies in Statistical Mechanics, Volume II*, (Amsterdam: North-Holland Publishing Company, 1964).
7. W. E. Morrell and J. H. Hilderbrand, The distribution of molecules in a model liquid, *J. Chem. Phys.*, **4**, 224–227, (1936).
8. J. G. Kirkwood, *Collected Papers*, ed. I. Oppenheim, (New York: Gordon and Breach Publishing, 1968).
9. J. G. Kirkwood, *Theory of Liquids*, ed. B. Alder, (New York: Gordon and Breach Publishing, 1968).

9.7 Exercises

1. Consider N classical particles in volume V at temperature T. Demonstrate that for the n-particle density, defined in Eq. 9.14, the following is true:

$$\int \cdots \int \rho_N^{(n)}(\underline{r}_1, \underline{r}_2, \ldots, \underline{r}_n) d\underline{r}_1 d\underline{r}_2 \ldots d\underline{r}_n = \frac{N!}{(N-n)!}. \qquad (9.78)$$

The density $\rho_N^{(n)}$ is called the n-particle density.

2. The mass density of liquid argon at $T = 143$ K is $\rho_m = 0.98 \times 10^3$ kg m^{-3}. Assume that the argon atoms interact via a Lennard-Jones potential with $\sigma = 3.4 \times 10^{-10}$ m and $\varepsilon = 1.66 \times 10^{-21}$ J.

   Determine the molar internal energy of liquid argon. How does the determined value compare to the observed one of -1.77×10^3 J mol^{-1}? Find missing information in the NIST Chemistry WebBook (*http://webbook.nist. gov/chemistry/fluid/*), or in the published literature.

3. Prove Eq. 9.43.

4. Find the relationship between the entropy S and the radial distribution function $g(r)$ for a non-ideal fluid, using the equation of state calculated with the radial distribution function, and the Maxwell relation for $\partial S/\partial V|_{N,T}$. What type of molecular level information will you need to determine the entropy quantitatively?

5. Determine the specific heat of a two-dimensional crystal according to the Einstein model.

6. Determine the heat capacity of silver as a function of the temperature, using the Einstein and the Debye models. Compare to literature values and comment on the temperature ranges of validity of these two models.

7. The room temperature heat capacity of beryllium is approximately 16.4 J/mol K and that of diamond is approximately 8.2 J/mol K. Explain why the Dulong–Petit model fails to predict the heat capacity of beryllium and diamond.

Beyond pure, single-component systems

10.1 Ideal mixtures

Consider an ideal gas consisting of N_1 particles of substance 1, N_2 particles of substance 2, ..., and N_K particles of substance K. The number K of different substances can be arbitrarily high.

The Hamiltonian of the system is

$$H(\underline{p}) = \sum_{k=1}^{K} \sum_{i=1}^{N_k} \frac{|\underline{p}_{i(k)}|^2}{2m_k}, \qquad (10.1)$$

where $\underline{p}_{i(k)} = (p_{i(k),x}, p_{i(k),y}, p_{i(k),z})$ is the momentum vector of particle i of component k, and m_k is its mass.

The total partition function is

$$Q(N, V, T) = \frac{V^N}{N_1! N_2! \ldots N_K! h^{3N}} \int d\underline{p} \exp\left(-\beta H(\underline{p})\right), \qquad (10.2)$$

where

$$N_1 + N_2 + \cdots + N_K = N \qquad (10.3)$$

and

$$\underline{p} = \left(\underline{p}_{1(1)}, \ldots, \underline{p}_{N_1(1)}, \underline{p}_{1(2)}, \ldots, \underline{p}_{N_2(2)}, \ldots, \underline{p}_{1(K)}, \ldots, \underline{p}_{N_K(K)}\right). \qquad (10.4)$$

With a quadratic term in the Hamiltonian for each momentum term, the integral in Eq. 10.2 can be determined to give

$$Q(N, V, T) = \frac{V^N}{\prod_{k=1}^{K} \left(N_k! \Lambda_k^{3N_k}\right)}, \qquad (10.5)$$

where Λ_k is the thermal wavelength of substance k.

The thermodynamic properties of the ideal mixture (Fig. 10.1) can now be determined. The Helmholtz free energy of the mixture is

$$A_{\text{mix}} = -k_B T \left\{ N \ln V - \sum_{k=1}^{K} \ln N_k! - \sum_{k=1}^{K} N_k \ln \Lambda_k^3 \right\}. \quad (10.6)$$

Using the particle density of each substance, $\rho_k = N_k/V$, and rearranging yields

$$A_{\text{mix}} = N k_B T \left\{ \sum_{k=1}^{K} x_k \ln \left(\rho_j \Lambda_k^3 \right) - 1 \right\}, \quad (10.7)$$

where $x_k = N_k/N$ is the molar fraction of component k.

The internal energy of an ideal gas mixture is again

$$E_{\text{mix}} = -\frac{\partial \ln Q}{\partial \beta} \bigg|_{N_1,\dots,N_K,V} = \frac{3 N k_B T}{2}. \quad (10.8)$$

The equation of state for the mixture is again determined as

$$P = k_B T \frac{\partial \ln Q}{\partial V} \bigg|_{N_1,\dots,N_K,T} = \frac{N k_B T}{V}. \quad (10.9)$$

Each component individually obeys the ideal gas law, i.e.,

$$x_k P V = x_k N k_B T = N_k k_B T. \quad (10.10)$$

Figure 10.1 Ideal mixing of two components.

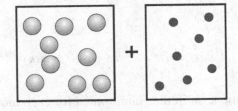

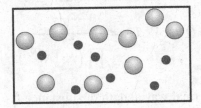

Let the partial pressure of each component be defined as

$$P_k = x_k P. \tag{10.11}$$

Then Eq. 10.10 yields Dalton's law

$$P_k V = N_k k_B T. \tag{10.12}$$

The entropy of the mixture is

$$S_{\text{mix}} = \frac{1}{T}(E_{\text{mix}} - A_{\text{mix}}). \tag{10.13}$$

Therefore

$$S_{\text{mix}} = Nk_B \left\{ \frac{5}{2} - \sum_{k=1}^{K} x_k \ln \left(\rho_k \Lambda_k^3 \right) \right\}. \tag{10.14}$$

The Gibbs free energy of the mixture is

$$G_{\text{mix}} = A_{\text{mix}} + PV. \tag{10.15}$$

This yields

$$G_{\text{mix}} = Nk_B T \sum_{k=1}^{K} x_k \ln \left(\rho_k \Lambda_k^3 \right). \tag{10.16}$$

Finally, the chemical potential of each substance in an ideal gas mixture is defined as

$$\mu_k^{IM} = \left. \frac{\partial G}{\partial N_k} \right|_{T,P,N_{j \neq k}} = k_B T \ln \left(\rho_k \Lambda_k^3 \right). \tag{10.17}$$

The chemical potential of any species k in an ideal gas mixture, μ_k^{IM}, is then related to the pure ideal gas component chemical potential, μ_k^{IG}, according to the following relation:

$$\mu_k^{IM} = \mu_k^{IG} + k_B T \ln x_k. \tag{10.18}$$

Clearly, whereas the pure component chemical potential depends only on the temperature and the pressure, the chemical potential of a mixture component also depends on the composition of the mixture.

The chemical potential μ_k^{IM} is also called the partial molar Gibbs free energy, denoted by $\overline{G}_k^{IM}$, which is related to the molar Gibbs free energy of the pure component by

$$\overline{G}_k^{IM} = \underline{G}_k^{IG} + k_B T \ln x_k. \tag{10.19}$$

For non-ideal mixtures, the chemical potential and the partial molar Gibbs free energy are not determined by as simple a relation as the one given in Eq. 10.18 or in Eq. 10.19. In engineering practice the concept of fugacity has proven useful in determining the deviation of real mixtures from the ideal mixture behavior. The fugacity of a component k in a mixture is generally defined as

$$\overline{f}_k(T, P, x_k) = x_k P \exp\left[\frac{\overline{G}_k(T, P, x_k) - \overline{G}_k^{IM}(T, P, x_k)}{k_B T}\right], \quad (10.20)$$

where $\overline{G}_k(T, P, x_k)$ is the actual chemical potential of component k in the mixture.

The fugacity of species can be determined experimentally in gaseous, liquid, and solid mixtures. The interested reader is referred to any engineering thermodynamics textbook for more details.

10.1.1 Properties of mixing for ideal mixtures

The change of entropy upon mixing is the difference between the entropy of the mixture and the sum of entropies of pure, individual components, that is

$$\Delta S_{\text{mix}} = S_{\text{mix}} - \sum_{k=1}^{K} S_k, \quad (10.21)$$

where S_k is the entropy of pure substance k.

Algebraic manipulation yields

$$\Delta S_{\text{mix}} = -N k_B \sum_{k=1}^{K} x_k \ln x_k. \quad (10.22)$$

In an analogous manner, the changes in the internal energy, enthalpy, Helmholtz free energy, and Gibbs free energy upon ideal mixing are found to be as follows:

$$\Delta U_{\text{mix}} = 0, \quad (10.23)$$

$$\Delta H_{\text{mix}} = 0, \quad (10.24)$$

$$\Delta A_{\text{mix}} = N k_B T \sum_{k=1}^{K} x_k \ln x_k, \quad (10.25)$$

and

$$\Delta G_{\text{mix}} = N k_B T \sum_{k=1}^{K} x_k \ln x_k, \qquad (10.26)$$

respectively.

The change in the volume upon mixing for an ideal mixture is

$$\Delta V_{\text{mix}} = 0. \qquad (10.27)$$

These are well-established thermodynamic relations that describe the properties of mixing as a function of the molar fractions of mixing components. Typically, these relations are presented in an ad hoc manner in thermodynamics textbooks. In this section, we have demonstrated that they can be naturally derived from first principles of statistical mechanics. Properties of non-ideal mixtures can also be derived, but, as expected, the derivations, which include particle interactions, quickly become cumbersome. The interested reader is referred to the literature at the end of the chapter for more details.

10.2 Phase behavior

Typical pressure–volume isotherms of real fluids are shown in Fig. 10.2. For temperatures higher than the critical temperature, the fluid exists in

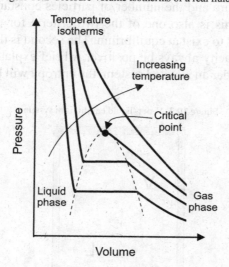

Figure 10.2 Pressure-volume isotherms of a real fluid.

a single homogeneous phase. Compressing the system decreases its volume, resulting in a monotonic increase of its pressure. The ideal gas law captures this behavior fairly accurately.

For temperatures below the critical pressure, there are ranges of values of the system volume where there are two distinct phases, a high-density liquid phase and a low-density gas phase, as illustrated in Fig. 10.3. How can this behavior be explained in terms of molecular interactions?

Consider a closed, single-component system described by the van der Waals (vdW) equation of state, which was introduced in Chapter 8. In Figure 8.4, the pressure is plotted as a function of the molar volume, $\underline{V} = (V/N)$, at various temperatures. These isotherms describe the fluid states of substances. What is remarkable about the van der Waals equation of state, for which van der Waals was awarded a Nobel prize in 1910, is that although very simple, it describes and predicts a first-order gas–liquid transition. In this section we discuss in detail this phase behavior of simple fluids.

At temperatures higher than a characteristic temperature, T_c, the critical temperature, the vdW pressure isotherms monotonically decrease with the molar volume. The slope of the curve is always negative

$$\left. \frac{\partial P}{\partial \underline{V}} \right|_T < 0. \tag{10.28}$$

This makes intuitive sense: as the volume of a system increases, keeping the temperature and the number of particles constant, the pressure will decrease. This is also one of the two criteria for thermodynamically stable states to exist at equilibrium. The second is that the constant volume heat capacity always be positive. A brief explanation is as follows: if we consider an isolated system, the entropy will be maximum at

Figure 10.3 Vapor–liquid equilibrium system.

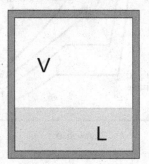

equilibrium, with $dS = 0$. The stability can be ensured if $d^2S < 0$. This means that the system will return to the same original equilibrium state if it is slightly perturbed away from this state. One can then show that this holds if $C_V > 0$ and Eq. 10.28 are both true (a detailed exposition of the criteria for equilibrium can be found in Sandler's *Chemical and Engineering Thermodynamics* textbook, see Further reading).

With the temperature and the pressure being the independent variables, at temperatures $T > T_c$ only a single, homogeneous, well-defined phase exists, because of the monotonic decrease of the pressure with increased volume. At any temperature lower than the critical temperature, and any given pressure, the van der Waals equation yields three values for the molecular volume. This is shown by expanding the vdW equation to obtain a cubic equation in volume:

$$V^3 - \left(Nb + \frac{Nk_BT}{P}\right)V^2 + \left(\frac{N^2\alpha^2}{P}\right)V + \left(\frac{N^2\alpha^2b}{P}\right) = 0. \quad (10.29)$$

This is also called a cubic equation of state. If solved for the molar volume, it predicts, for certain temperatures and pressures, three distinct phases with well-defined molar volumes, $\underline{V}_1, \underline{V}_2, \underline{V}_3$.

It is important to note, however, that the phase of the intermediate molar volume, $\underline{V}_2$ is not thermodynamically stable. This is because, according to the vdW equation,

$$\left.\frac{\partial P}{\partial \underline{V}}\right|_{T, \underline{V} = \underline{V}_2} > 0. \quad (10.30)$$

In reality only two separate, distinct phases with $\underline{V}_1$ and $\underline{V}_3$ appear below the critical temperature. These are the liquid and gas phases, with the smaller and larger values of the molar volume, respectively. The equilibrium conditions of constant temperature, pressure, and chemical potential are valid for each of these phases.

Consider now an isothermal compression process: starting with N particles in the gas phase, compress the volume and increase the pressure keeping the temperature constant. When the molecular volume reaches the value $\underline{V}_3$, the first droplet of liquid will appear in the system with density $1/\underline{V}_1$. As the volume is compressed further, the pressure of the system does not change. Instead, more liquid is formed, lowering the average molecular volume of the system. This constant system pressure is the vapor pressure of the system at the given temperature. At a certain point, when $V/N = \underline{V}_1$, all of the gas will be liquefied. If the volume is compressed further the pressure will increase sharply because of the small compressibility of liquids. This is the behavior observed for temperatures lower than the critical temperature.

At the critical point there is no distinction between liquid and gas phases. Both derivatives of the pressure with respect to the volume vanish at this point, i.e.,

$$\left.\frac{\partial P}{\partial \underline{V}}\right|_{T=T_c} = 0 \tag{10.31}$$

and

$$\left.\frac{\partial^2 P}{\partial \underline{V}^2}\right|_{T=T_c} = 0. \tag{10.32}$$

Using the vdW equation of state and these critical point conditions we can show that

$$T_C = \frac{8a}{27Rb}, \tag{10.33}$$

$$P_C = \frac{a}{27b^2}, \tag{10.34}$$

and

$$V_C = 3b. \tag{10.35}$$

Remembering that

$$\begin{cases} a = \frac{2\pi}{3}\sigma^3\varepsilon \\ b = \frac{2\pi}{3}\sigma^3 \end{cases} \tag{10.36}$$

we note that molecular parameters are sufficient to determine the critical point.

Importantly, for all vdW gases the critical point satisfies the universal relation

$$\frac{P_C V_C}{N k_B T_C} = 0.375. \tag{10.37}$$

In experiments, this value varies typically between 0.25 and 0.4. This confirms that the vdW equation is a particularly useful equation of state, albeit not absolutely accurate. Other cubic equations have been developed that more accurately capture the fluid states of substances, such as the Redlich–Kwong and the Peng–Robinson equations of state.

The interested reader is referred to standard thermodynamics textbooks for more details.

10.2.1 The law of corresponding states

Remember from Chapter 8 that the pressure of monoatomic NVT-systems is given by

$$P = k_B T \frac{\partial \ln Z}{\partial V}\bigg|_{N,T}, \tag{10.38}$$

where $Z(N, V, T)$ is the configurational integral.

Assuming only pairwise interactions,

$$Z(N, V, T) = \int d\underline{r} \exp(-\beta U(\underline{r})), \tag{10.39}$$

where

$$U(\underline{r}) = \sum_{ij} u(\underline{r}_{ij}) \tag{10.40}$$

is the sum of all pairwise interaction potentials.

Introducing the dimensionless variables

$$
\begin{aligned}
r^* &= r/\sigma, \\
u^*(r^*) &= u(r)/\varepsilon, \\
T^* &= k_B T/\varepsilon, \\
V^* &= V/\sigma^3, \\
\rho^* &= N/V^*, \\
P^* &= P\sigma^3/\varepsilon, \\
Z^* &= Z/\sigma^{3N},
\end{aligned}
\tag{10.41}
$$

we obtain

$$P^* = k_B T^* \frac{\partial \ln Z^*}{\partial V^*}\bigg|_{N,T^*}. \tag{10.42}$$

Equation 10.42 implies that for fluids for which the pairwise potential has the same functional form (for example, all Lennard-Jones fluids), the pressure is a universal function of the reduced temperature T^* and number density ρ^*. In other words, there is a single, universal P^*V^* diagram that describes the behavior of fluids with the same pair potential function. This is the law of corresponding states, which holds for classes of structurally similar chemical substances.

A dimensionless critical point, which is the same for all fluids, can be defined with

$$P_C^* = P_C \sigma^3 / \varepsilon,$$
$$V_C^* = V_C / \sigma^3,$$
$$T_C^* = k_B T_C / \varepsilon. \tag{10.43}$$

We can further define reduced thermodynamic quantities as

$$P_r = P/P_C, \quad V_r = V/V_C, \quad T_r = T/T_C. \tag{10.44}$$

Then the macroscopic law of corresponding states for all fluids can be written as

$$P_r = \frac{P^*(V_C^* V_r, T_C^* T_r)}{P_C^*}. \tag{10.45}$$

A compressibility factor can be defined that can capture corresponding states equations:

$$z_r = \frac{P_r V_r}{T_r}. \tag{10.46}$$

In Fig. 10.4, the compressibility factor of common gases is shown validating the conformity of a variety of fluids to the law of corresponding states.

10.3 Regular solution theory

Regular solution theory employs simple statistical arguments to capture the nature of phase equilibria of solutions. Although simple, it has been used widely to predict the existence of phases and phase transitions. It has been used to explain different first-order transition phenomena, such as ferromagnetism and gas adsorption on solid surfaces.

Consider a system of N_1 molecules of species 1 and N_2 molecules of species 2, as illustrated in Fig. 10.5. The species can be different chemical moieties or different phases of the same substance. Assume that the molecules are positioned at the $N = N_1 + N_2$ sites of a regular three-dimensional lattice. Each lattice site is assumed to occupy the same volume v_o, so that the total volume of the system is $V = N v_o$. The coordination number z is the number of sites neighboring each lattice site. Each molecule is assumed to interact only with its z nearest neighbors.

Figure 10.4 Compressibility factors for different fluids. Reprinted with permission from Gour-Jen Su, *Ind. Eng. Chem.*, **38**, (1946), p. 803, © American Chemical Society.

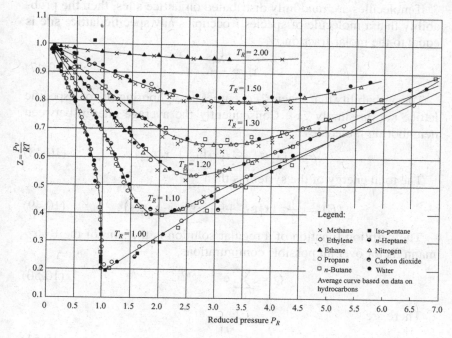

Figure 10.5 Regular solution lattice.

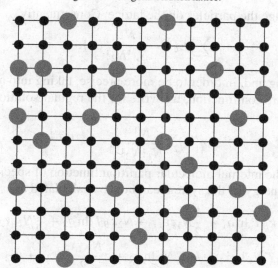

The potential energy of interaction is ε_{11}, ε_{22}, and ε_{12} between molecules in neighboring sites, depending on the type of molecules.

If molecules are randomly distributed on lattice sites, then the probability that a molecule of species i occupies any specific lattice site is equal to the molar fraction x_i,

$$x_i = \frac{N_i}{N}. \tag{10.47}$$

The probability that species j occupies a neighboring site of a specific lattice site is simply zx_j. Consequently the average energy between neighboring molecules is

$$\langle \varepsilon \rangle = z \left(x_1^2 \varepsilon_{11} + x_1 x_2 \varepsilon_{12} + x_2^2 \varepsilon_{22} \right). \tag{10.48}$$

The total energy of the system is then

$$\langle E \rangle = \frac{N}{2} z \left(x_1^2 \varepsilon_{11} + x_1 x_2 \varepsilon_{12} + x_2^2 \varepsilon_{22} \right). \tag{10.49}$$

The partition function of a regular solution is the sum of the Boltzmann factor over all possible configurations,

$$Q = \sum_C e^{-E_C/k_B T}. \tag{10.50}$$

There are

$$\frac{N!}{N_1! N_2!} \tag{10.51}$$

ways to arrange the particles on the lattice. Consequently,

$$\sum_C E_C = \frac{N!}{N_1! N_2!} \langle E \rangle. \tag{10.52}$$

Simplifying and, in order to be more precise, taking into account the molecular partition function, we write for the regular solution partition function

$$Q = q_1^{N_1} q_2^{N_2} \frac{N!}{N_1! N_2!} e^{-\langle E \rangle / k_B T}, \tag{10.53}$$

where q_i is the internal molecular partition function of species i.

The free energy of the system can now be calculated as

$$A = -N_1 k_B T \ln q_1 + \frac{1}{2} N_1 z \varepsilon_{11} - N_2 k_B T \ln q_2 + \frac{1}{2} N_2 z \varepsilon_{22}$$

$$+ k_B T N_1 \ln x_1 + k_B T N_2 \ln x_2 + z \frac{N x_1 x_2}{2} (\varepsilon_{11} + \varepsilon_{22} - 2\varepsilon_{12}). \tag{10.54}$$

The chemical potential of either species is then

$$\mu_i = -k_B T \ln q_i + \frac{1}{2} z \varepsilon_{ii} + k_B T \ln x_i - \frac{z}{2}(1 - x_i)^2(\varepsilon_{11} + \varepsilon_{22} - 2\varepsilon_{12}).$$
(10.55)

With these equations, regular solution theory provides the tools to determine vapor–liquid equilibria of binary mixtures, which we briefly present in the following section. The interested reader can consult the thoroughly excellent book *Statistical Mechanics of Phases, Interfaces, and Thin Films* by H. Ted Davis (see Further reading).

10.3.1 Binary vapor–liquid equilibria

Consider a binary liquid solution at equilibrium with its vapor.

The chemical potentials of the two species in the liquid phase, μ_i^L, can be described by Eq. 10.55. On the other hand, the chemical potential of each species in the gas phase is

$$\mu_i^G = -k_B T \ln q_i + k_B T \ln P_i,$$
(10.56)

where $P_i = y_i P$ is the partial pressure of species i.

At equilibrium

$$\mu_i^G = \mu_i^L.$$
(10.57)

Thus,

$$k_B T \ln P_i = \frac{1}{2} z \varepsilon_{ii} - \frac{z}{2}(1 - x_i)^2(\varepsilon_{11} + \varepsilon_{22} - 2\varepsilon_{12}).$$
(10.58)

From Eq. 10.58, the vapor pressure of pure component i, $P_i^V(T)$, is

$$P_i^V(T) = \exp\left(\frac{z \varepsilon_{ii}}{2 k_B T}\right).$$
(10.59)

Rearranging Eq. 10.58 and using Eq. 10.59 results in the following expression for binary vapor–liquid equilibria:

$$\frac{y_i P}{x_i P_i^V(T)} = \exp\left(-\frac{z}{2 k_B T}(1 - x_i)^2(\varepsilon_{11} + \varepsilon_{22} - 2\varepsilon_{12})\right).$$
(10.60)

For an ideal solution with no interactions, Eq. 10.60 reduces to the well-known Raoult's law:

$$y_i P = x_i P_i^V(T).$$
(10.61)

Equation 10.60 illustrates the source of deviation from Raoult's law for simple binary solutions. Depending on the nature of the particle–particle interactions there may be positive or negative deviations from Raoult's law. In particular, in negative deviations, species 1 and 2 are attracted more strongly to one another than to their own kind. The solution vapor pressure is then lower than the one calculated by Raoult's law. The opposite occurs in positive deviations.

10.4 Chemical reaction equilibria

Consider the reversible chemical reaction

$$\nu_A A + \nu_B B \underset{k_2}{\overset{k_1}{\rightleftharpoons}} \nu_C C + \nu_D D, \tag{10.62}$$

where $\nu_i, i = A, B, C, D$, are the stoichiometric coefficients of the reacting molecules, and k_1, k_2 are the forward and reverse kinetic constants, respectively.

Assume that the system has a constant number of molecules $N = \sum_i N_i$ and is at constant temperature and constant volume. It follows that at equilibrium the free energy A is minimum, or

$$dA = -SdT - PdV + \sum_i \mu_i dN_i = 0. \tag{10.63}$$

Since the temperature and the volume of the system are constant, at equilibrium

$$\sum_i \mu_i dN_i = 0. \tag{10.64}$$

Changes in the numbers of molecules because of the reaction are not independent. Instead they are related as follows:

$$dN_A = \nu_A d\lambda,$$
$$dN_B = \nu_B d\lambda,$$
$$dN_C = \nu_C d\lambda,$$
$$dN_D = \nu_D d\lambda, \tag{10.65}$$

where $d\lambda$ is an arbitrary change in the number of molecules according to Eq. 10.62.

Consequently, the criterion for chemical equilibrium for a single reaction is

$$\sum_i \nu_i \mu_i = 0. \tag{10.66}$$

This criterion can be extended to many reaction systems. If there are M reactions, then at equilibrium

$$\sum_i \nu_{ij} \mu_i = 0, \quad j = 1, \ldots, M, \tag{10.67}$$

where ν_{ij} is the stoichiometric coefficient of molecular species i in reaction j.

For a single reaction, the equilibrium constant is defined as

$$K_{eq} = \frac{(N_C/V)^{\nu_C}(N_D/V)^{\nu_D}}{(N_A/V)^{\nu_A}(N_B/V)^{\nu_B}}, \tag{10.68}$$

which reduces to

$$K_{eq} = \frac{N_C^{\nu_C} N_D^{\nu_D}}{N_A^{\nu_A} N_B^{\nu_B}}, \tag{10.69}$$

because $\sum_i \nu_i = 0$.

If the reaction occurs in an ideal gas mixture, the equilibrium constant can be easily determined with statistical arguments. The partition function of the system is

$$Q(N, V, T) = \frac{q_A^{N_A} q_B^{N_B} q_C^{N_C} q_D^{N_D}}{N_A! \, N_B! \, N_C! \, N_D!}, \tag{10.70}$$

where q_i is the canonical partition function of each molecule of chemical species i.

The chemical potential μ_i is defined as

$$\mu_i = \left. \frac{\partial A}{\partial N_i} \right|_{V,T,N_{j \neq i}}$$

$$= -k_B T \left. \frac{\partial \ln Q}{\partial N_i} \right|_{V,T,N_{j \neq i}}. \tag{10.71}$$

We can then write for the chemical potential, using Eq. 10.70,

$$\mu_i = -k_B T \ln \frac{q_i}{N_i}. \tag{10.72}$$

At equilibrium we find, using Eq. 10.72 and Eq. 10.67,

$$\frac{N_C^{\nu_C} N_D^{\nu_D}}{N_A^{\nu_A} N_B^{\nu_B}} = \frac{q_C^{\nu_C} q_D^{\nu_D}}{q_A^{\nu_A} q_B^{\nu_B}}. \tag{10.73}$$

Consequently, the equilibrium constant can be computed from the molecular partition functions as follows:

$$K_{eq}(T) = \frac{q_C^{v_C} q_D^{v_D}}{q_A^{v_A} q_B^{v_B}}.$$ (10.74)

The chemical potential is also defined as

$$\mu_i = \mu_i^o(T) + k_B T \ln P_i,$$ (10.75)

where $\mu_i^o(T)$ is the standard chemical potential and P_i is the partial pressure of component i.

We can define a different equilibrium constant as follows:

$$K_P(T) = (k_B T)^{\sum_i v_i} K_{eq}(T),$$ (10.76)

or, using Eq. 10.76

$$K_P(T) = \exp\left(-\frac{\Delta A^o}{k_B T}\right),$$ (10.77)

where $\Delta A^o = \sum_i v_i \mu_i^o$ is the standard free energy change for the reaction.

10.5 Further reading

1. H. T. Davis, *Statistical Mechanics of Phases, Interfaces, and Thin Films*, (New York: VCH Publishers, 1996).
2. D. A. McQuarrie and J. D. Simon, *Physical Chemistry: A Molecular Approach*, (Sausalito, CA: University Science Books, 1997).
3. J. H. Hilderbrand, J. M. Prausnitz, and R. L Scott, *Regular and Related Solutions*, (Princeton, NJ: Van Nostrand-Reinhold, 1970).
4. S. I. Sandler, *Chemical and Engineering Thermodynamics*, (New York: Wiley, 1999).

10.6 Exercises

1. Prove 10.22, showing that it implies isobaric, isothermal mixing of ideal gases.

2. Prove 10.18.

3. Starting with Eqs. 10.31 and 10.32, prove relations 10.33–10.35.

4. The critical point temperature and pressure of oxygen are $T_C = 154.59$ K and $P_C = 5.043$ MPa, respectively. Estimate the Lennard-Jones parameters of oxygen. Calculate the critical point molar volume and compare to the experimental value (you can find it in the NIST Chemistry WebBook , at http://webbook.nist.gov/chemistry/).

5. Using the Lennard-Jones parameters, determine the second virial coefficient of oxygen at $T = 100\,\text{K}$, $T = 200\,\text{K}$ and $T = 300\,\text{K}$. Assume a concentration of 0.1 M. Determine the pressure of oxygen using the ideal gas law and the virial equation of state. Compare to experimental values (you can find these in the NIST Chemistry WebBook, at http://webbook.nist.gov/chemistry/).

6. The critical temperature and critical pressure of propane are $T_C = 370\,\text{K}$ and $P_C = 42\,\text{atm}$, respectively. Assume that the van der Waals equation of state describes the behavior of propane well. Determine the Lennard-Jones parameters, the critical molar volume, the molar entropy, the molar energy, and the heat of vaporization at 1 atm.

11

Polymers – Brownian dynamics

Polymers are remarkable molecules with a particularly rich behavior and a wealth of interesting properties. Statistical mechanical arguments may be used to understand these properties. In this chapter, we present an elementary theory of polymer configurations and polymer dynamics. We also offer a brief exposition of the theory of Brownian dynamics. This is a powerful theoretical formalism for studying the motion of molecules in a solution.

11.1 Polymers

The study of conformations and conformational motions of flexible polymer chains in solution is of great scientific and technological importance. Understanding the physics of macromolecules at the molecular level helps the synthesis and design of commercial products. It also provides insight into the structure and functions of biological systems. Flexible polymers have therefore been the subject of extensive theoretical treatments, a wide variety of experiments, and computer simulations (see Further reading at the end of the chapter).

Historically, theoretical treatments have resorted to simple phenomenological models of polymeric materials. In the framework of statistical mechanics, polymeric chains are at a first stage considered to consist of independent elements or segments. The principal property of macromolecular behavior taken into account with this representation is the flexibility of the chains. With non-interacting monomeric units having uncorrelated directions, it is straightforward to show that the chains acquire random-walk behavior.

In what follows, we present this elementary theory of polymer conformations and of polymer dynamics.

11.1.1 Macromolecular dimensions

In this section, we present statistical arguments for determining the size of polymer chains. Consider as an example a polyethylene chain

(Fig. 11.1). In the most extended conformation, the maximum length of the chain is determined by the number of monomers M, the length of the bond between monomers l, and the bond angle θ,

$$L_{\max} = (M - 1)l \sin(\theta/2). \qquad (11.1)$$

Chains can rotate around dihedral angles. Inevitably then, the average length of polyethylene chains will be smaller than $L_{\max}$.

The simplest model to determine the average length $\langle L \rangle$ of chains with M monomers is the freely jointed ideal chain. In a freely jointed chain, the dihedral angles can take on any value with equal probability. In an ideal chain there are no interactions between the monomers. In other words, the polymer is assumed to be a random walk in three dimensions (Fig. 11.2). Lord Rayleigh developed the model of unrestricted random walks in the early 1900s and Werner Kuhn first adopted this model in 1934 for polymer conformations.

Figure 11.1 Polyethylene monomer.

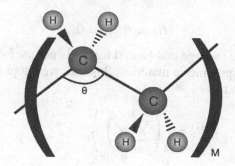

Figure 11.2 Random polymer conformation.

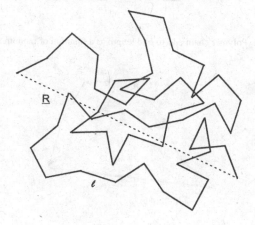

The polymer length (Fig. 11.3) is simply the vectorial sum of all bond length vectors projected on the dimension of the end-to-end vector of the chain:

$$\langle L \rangle = \langle |\underline{R}| \rangle. \tag{11.2}$$

If we denote with $\underline{x}_i$ the position of monomer i, and with $\underline{u}_i = \underline{x}_{i+1} - \underline{x}_i$ the bond vector between monomers, we can write

$$\underline{R} = \sum_{i=1}^{M-1} \underline{u}_i$$

$$= \sum_{i=1}^{M-1} l \cos(\phi), \tag{11.3}$$

where $l = |\underline{u}_i|$ is the bond length, and ϕ is the angle between the bond vector and the polymer end-to-end vector.

Since the chain is freely jointed, we have $\langle \cos(\phi) \rangle = 0$, assuming $M \to \infty$. Then

$$\langle L \rangle = \langle |\underline{R}| \rangle = 0. \tag{11.4}$$

Obviously, the average end-to-end length is not useful for characterizing the size of polymeric materials. Instead, we adopt the root-mean-square end-to-end size

$$\langle \underline{R}^2 \rangle = \left\langle \left(\sum_{i=1}^{M-1} \underline{u}_i \right)^2 \right\rangle$$

$$= \sum_{i=1}^{M-1} \langle \underline{u}_i^2 \rangle + 2 \sum_{i \neq j}^{m-1} \langle \underline{u}_i \cdot \underline{u}_j \rangle, \tag{11.5}$$

Figure 11.3 Polymer chain end-to-end length as a function of monomer bonds.

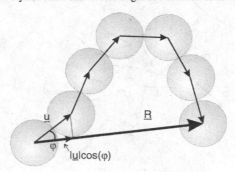

where

$$\langle \underline{u}_i^2 \rangle = l^2. \tag{11.6}$$

For an ideal freely jointed chain the angle θ_{ij} between two vectors $\underline{u}_i$ and $\underline{u}_j$ can take any value between 0 and 2π with equal probability. Then

$$\langle \underline{u}_i \cdot \underline{u}_j \rangle = l^2 \langle \cos(\theta_{ij}) \rangle = 0. \tag{11.7}$$

Consequently,

$$\langle \underline{R}^2 \rangle = M l^2. \tag{11.8}$$

The mean size of a macromolecule can be defined as the root-mean-square end-to-end distance

$$R = \langle \underline{R}^2 \rangle^{1/2} = M^{0.5} l. \tag{11.9}$$

The exponent 0.5 appearing in Eq. 11.9 is called the scaling exponent of an ideal chain.

Since the polymer can be considered as a random walk, the statistical distribution of R is Gaussian:

$$P_M(R) = \left(\frac{3}{2\pi M l^2} \right)^{3/2} \exp \left(\frac{-3R^2}{2M l^2} \right). \tag{11.10}$$

Therefore,

$$\langle \underline{R}^2 \rangle = \int \underline{R}^2 P_M(R) dR. \tag{11.11}$$

We can prove this, starting with the joint probability of all monomer positions

$$P(\underline{x}_1, \underline{x}_2, \ldots, \underline{x}_M). \tag{11.12}$$

The positions of monomers are not independent, but in a freely jointed chain the position of the $i + 1$ monomer depends only on the position of the i monomer. Let us denote with $p(\underline{x}_{i+1} | \underline{x}_i)$ the conditional probability of finding the $i + 1$ monomer at position $\underline{x}_{i+1}$ provided that the i monomer is at position $\underline{x}_i$.

We can write

$$P(\underline{x}_1, \underline{x}_2, \ldots, \underline{x}_M) = p(\underline{x}_2 | \underline{x}_1) p(\underline{x}_3 | \underline{x}_2) \ldots p(\underline{x}_M | \underline{x}_{M-1}). \tag{11.13}$$

The functional form of the conditional probability is

$$p(\underline{x}' | \underline{x}) = \frac{1}{4\pi l^2} \delta \left(|\underline{x} - \underline{x}'| - l \right). \tag{11.14}$$

In other words, $\underline{x}'$ can be anywhere on a sphere of radius l centered at $\underline{x}$ (Fig. 11.4).

We can write

$$p(\underline{x}'|\underline{x}) = p(\underline{x}' - \underline{x}) = p(\underline{u}_i). \tag{11.15}$$

Consequently

$$P_M(R) = \int P(\underline{x}_1, \underline{x}_2, \ldots, \underline{x}_M; R = |\underline{x}_M - \underline{x}_1|)d\underline{x}_1 \ldots d\underline{x}_M, \tag{11.16}$$

which yields

$$P_M(R) = \int \delta\left(\left|\sum_{i=1}^{M-1} \underline{u}_i - \underline{R}\right|\right) \prod_{i=1}^{M-1} p(\underline{u}_i)d\underline{u}_1 \ldots d\underline{u}_{M-1}. \tag{11.17}$$

After considerable algebraic manipulations, similar to the central limit theorem derivation in Chapter 2, the result is the Gaussian distribution in Eq. 11.10.

11.1.2 Rubber elasticity

Elasticity is the property of materials that can deform reversibly under external forces, or stress. The relative deformation amount is called the strain. For example, Hooke's spring has linear elasticity with the force being proportional to the deformation, $F = kx$, where k is the spring constant.

Thomas Young introduced a property of materials, which is now known as the Young modulus or the elasticity modulus, that quantifies their elastic behavior. The Young modulus, E, is defined as

$$E = \frac{\sigma}{\varepsilon}. \tag{11.18}$$

Figure 11.4 Random positions of consecutive monomers.

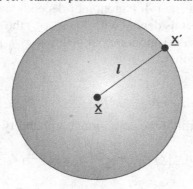

In Eq. 11.18, σ is the tensile stress, defined as the applied force, F, over the area, A,

$$\sigma = \frac{F}{A}. \tag{11.19}$$

In Eq. 11.18, ε is the tensile strain, defined as the ratio of the amount of deformation ΔL over the original length, L (see Fig. 11.5)

$$\varepsilon = \frac{\Delta L}{L}. \tag{11.20}$$

Many polymers, such as rubber, are characterized by high elasticity. In the 1950s, Volkestein in Leningrad and Paul Flory at Cornell worked out the details that connect the strain of a polymer to applied stress. Using statistical arguments they showed how the two are linearly related, closely following Hooke's law.

Volkestein started his analysis by focusing on the probability distribution of the size of polymer chains. He demonstrated that this probability distribution is determined as the ratio of two configurational integrals,

$$P_M(R) = \frac{Z_R}{Z}, \tag{11.21}$$

where Z is the configurational integral of a chain with M monomers of arbitrary size and Z_R is the configurational integral of a chain with M monomers of size R.

The configurational or, in other words, the elastic free energy of a chain of size R is defined as

$$\begin{aligned} A_R &= -k_B T \ln Z_R \\ &= C - k_B T \ln P_M(R), \end{aligned} \tag{11.22}$$

where the constant $C = C(T)$ is only dependent on the temperature.

Figure 11.5 Material deformation ΔL as a result of applied tensile force F.

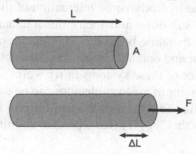

Assuming a Gaussian distribution for the polymer size, as in Eq. 11.10, the free energy becomes

$$A_R = C(T) + \left(\frac{3k_B T}{2Ml^2}\right) R^2. \qquad (11.23)$$

Differentiating the elastic free energy with respect to R yields the average force exerted by a chain of size R,

$$f_R = \left(\frac{3k_B T}{2Ml^2}\right) R. \qquad (11.24)$$

The term

$$\left(\frac{3k_B T}{2Ml^2}\right) \qquad (11.25)$$

is the elastic coefficient, or Young modulus, of a single chain.

Flory reached the same result and premised that the elastic energy of plastic materials resides within the chains, and that the stress can be determined as a sum of terms similar to the terms of Eq. 11.24. He was thus able to connect the macroscopic strain of a polymeric material to the conformations of individual chains. For his work, Paul Flory was awarded the 1974 Nobel Prize in Chemistry.

11.1.3 Dynamic models of macromolecules

Features of the dynamic behavior of polymers have been explained based on the idea of Brownian motion of the monomeric segments, which we detail in the following section. Here we present briefly the prominent treatises of conformational motions of flexible polymers in dilute solutions. These are the Rouse and the Zimm models.

The Rouse model assumes unperturbed chains in the free-draining limit (the limit of no hydrodynamic interactions between monomers) that can be represented as a set of beads connected along the chain contour. The Zimm model, in addition to the Rouse model features, takes into account the hydrodynamic interaction of the beads.

For concentrated solutions and melts the interactions of different chains take over the dynamic behavior. The reptation theory, based on the ideas of reptation and entanglements, first discussed by de Gennes, describes the behavior of these systems fairly well.

The dynamic behavior of macromolecules can be explained by adopting the time correlation functions of the equilibrium fluctuations of the chain end-to-end vectors, $c_{RR}(t)$ (for a precise definition of correlation functions, the reader should consult Chapter 12).

$$
c_{RR}(t) = \frac{\left\langle \sum_{i=1}^{M_c} \underline{R}_i(0) \cdot \sum_{k=1}^{M_c} \underline{R}_k(t) \right\rangle - \left\langle \sum_{i=1}^{M_c} \underline{R}_i \right\rangle^2}{\left\langle \sum_{i=1}^{M_c} \underline{R}_i(0) \cdot \sum_{k=1}^{M_c} \underline{R}_k(0) \right\rangle - \left\langle \sum_{i=1}^{M_c} \underline{R}_i \right\rangle^2}, \tag{11.26}
$$

where M_c is the number of polymer chains.

For systems in homogeneous environments and in the absence of cross-correlations between different chains, $c_{RR}(t)$ is reduced to the autocorrelation function of the end-to-end vector,

$$
c_{RR}(t) = \frac{\langle \underline{R}(0) \cdot \underline{R}(t) \rangle}{\langle \underline{R}(0) \cdot \underline{R}(0) \rangle}. \tag{11.27}
$$

Both the Rouse and the Zimm models, following a normal-mode analysis, propose the same form for this autocorrelation function,

$$
c_{RR}(t) = \frac{8}{\pi^2} \sum_{p=1}^{\infty} \frac{1}{p^2} \exp(-t/\tau_p), \quad (p : \text{odd}) \tag{11.28}
$$

where τ_p is the relaxation time of the pth eigenmode.

The difference between the two models lies in the scaling of relaxation times with the number of the modes. The Rouse model predicts

$$
\tau_{p\,\text{Rouse}} = \frac{\zeta M^2 b^2}{3\pi^2 k_B T p^2}, \quad (p : \text{odd}) \tag{11.29}
$$

where M is the number of monomer beads per chain, b is the average length between beads, and ζ is the friction coefficient of a bead.

The Zimm model predicts

$$
\tau_{p\,\text{Zimm}} = \pi^{3/2} \eta_s b^3 M^{3/2} / (12^{1/2} k_B T \lambda_p), \tag{11.30}
$$

where η_s is the solvent viscosity and λ_p is the pth eigenvalue (the first three are determined as $\lambda_1 = 4.04$, $\lambda_3 = 24.2$, $\lambda_5 = 53.3$).

The reptation theory, which postulates that entangled chains diffuse along their own contour inside a tube formed by the surrounding matrix of other macromolecules, predicts the same form for the correlation function with significantly longer correlation times:

$$
\tau_{p\,\text{rept}} = L^2 / \pi^2 p^2 D_c, \quad (p : \text{odd}) \tag{11.31}
$$

where L is the length of the tube in which the polymer chain is assumed to reptate and D_c is the diffusion coefficient of the chain along the tube.

The interested reader is referred to the rich polymeric materials literature for more details on the conformation and dynamics of polymers.

11.2 Brownian dynamics

Robert Brown's observation that small pollen grains are in perpetual irregular motion on the surface of water laid dormant for nearly eight decades, before Einstein explained it as the result of particle collisions with solvent molecules (Fig. 11.6). Of course, even Einstein was not entirely certain that what he was modeling was what Brown had observed. In 1905, Einstein wrote "It is possible that the motions to be discussed here are identical with the so-called Brownian molecular motion; however, the data available to me on the latter are so imprecise that I could not form a judgment on the question." It was actually Jean Perrin who accurately measured Brownian motion and connected Einstein's theory to Brown's observations in 1909.

Einstein's theory of Brownian motion was ground-breaking because it substantiated the atomic hypothesis, or the "Atoms Doctrine," as Perrin calls it in his wonderful book. First enunciated by the ancient Greeks, the idea that all matter consists of minute, indivisible entities called atoms became a sound scientific hypothesis with the work of Dalton, Boltzmann, and Mendeleyev. But it was not until Einstein worked on the

Figure 11.6 Brownian particles experience a large number of collisions from solvent molecules.

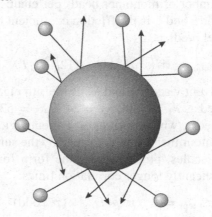

theory of Brownian motion that a theory with sufficient empirical content was devised that unambiguously supported the atomic hypothesis.

Einstein determined, using osmotic pressure arguments, that the mean-squared displacement of a particle from an original position $\underline{r}_0$ is proportional to time,

$$\langle |\underline{r} - \underline{r}_0|^2 \rangle = \frac{RT}{N_A \pi \alpha \mu} t, \tag{11.32}$$

where α is the size of the particle, and μ is the solvent viscosity.

It was three years later, in 1908, that Paul Langevin devised an alternative theory that gives the same correct result as Einstein's. Because Langevin's theory is applicable to a large class of random processes, it is the one predominantly discussed to explain Brownian motion. Here we describe its important elements. Langevin devised the following equation that describes the Brownian motion of a particle of mass M:

$$M \frac{dV}{dt} = -\gamma \underline{V} + \underline{F}'_s, \tag{11.33}$$

where $\underline{V}$ is the velocity of the particle, γ is the friction coefficient, and $\underline{F}'_s$ is a fluctuating force acting on the particles as a result of rapidly colliding solvent molecules.

The force is independent of the velocity of the particle and has the following properties:

$$\langle \underline{F}'_s(t) \rangle = 0 \tag{11.34}$$

and

$$\langle \underline{F}'_s(t) \cdot \underline{F}'_s(t + \tau) \rangle \neq 0. \quad \text{only for } \tau \to 0 \tag{11.35}$$

Dividing Eq. 11.33 by M yields

$$\frac{dV}{dt} = -\zeta \underline{V} + \underline{F}_s, \tag{11.36}$$

where $\zeta = \gamma/M$ and $\underline{F}_s = \underline{F}'_s/M$.

Solving Eq. 11.36 with initial velocity $\underline{V}_o$ yields

$$\underline{V}(t) = \underline{V}_o e^{-\zeta t} + e^{-\zeta t} \int_0^t e^{\zeta t'} \underline{F}_s(t') dt'. \tag{11.37}$$

Because of Eq. 11.34, ensemble averaging both sides of Eq. 11.37 yields

$$\langle \underline{V} \rangle = \underline{V}_o e^{-\zeta t}. \tag{11.38}$$

The velocity mean-squared average can be determined by squaring both sides in Eq. 11.37 and then taking the ensemble average,

$$\langle \underline{V}^2 \rangle = \underline{V}_o^2 e^{-2\zeta t} + e^{-2\zeta t} \int_0^t \int_0^t e^{\zeta(t'+t'')} \langle \underline{F}_s(t') \cdot \underline{F}_s(t'') \rangle \, dt' dt'',$$

(11.39)

where

$$\langle \underline{F}_s(t') \cdot \underline{F}_s(t'') \rangle \neq 0. \quad \text{only for } |t' - t''| \to 0$$

(11.40)

Consequently,

$$\langle \underline{V}^2 \rangle = \frac{3k_B T}{m} + \left(V_o^2 - \frac{3k_B T}{m} \right) e^{-2\zeta t}.$$

(11.41)

The displacement of the particle can be determined as

$$\underline{r} - \underline{r}_o = \int_0^t \underline{V}(t') dt',$$

(11.42)

and averaging

$$\langle \underline{r} - \underline{r}_o \rangle = \underline{V}_o \frac{1 - e^{-\zeta t}}{\zeta}.$$

(11.43)

Therefore,

$$\langle |\underline{r} - \underline{r}_o|^2 \rangle = \frac{V_o^2}{\zeta^2} \left(1 - e^{-\zeta t} \right)^2 + \frac{3k_B T}{M\zeta^2} \left(2\zeta t - 3 + 4e^{-\zeta t} - e^{-2\zeta t} \right).$$

(11.44)

At the limit of $t \to \infty$

$$\langle |\underline{r} - \underline{r}_o|^2 \rangle = \frac{6k_B T}{M\zeta} t,$$

(11.45)

which is, according to Einstein, equal to $6Dt$, where D is the particle diffusion coefficient

$$D = \frac{k_B T}{M\zeta}.$$

(11.46)

The significance of Langevin's and Einstein's contributions lies with the calculation of the displacement of Brownian particles. Whereas scientists before them were after the particle "velocity of agitation," Einstein developed a sound theory for a quantity that could be measured with high precision, the particle diffusion coefficient. Langevin captured this in a generalized model and Perrin unambiguously validated

Einstein's predictions by simply verifying that the mean displacement of a particle is doubled when the time is increased fourfold.

11.3 Further reading

1. P. G. de Gennes, *Scaling Concepts in Polymer Physics*, (Cornell, NY: Cornell University Press, 1979).
2. M. Doi and S. F. Edwards, *The Theory of Polymer Dynamics*, (London: Oxford Science Publishing, 1986).
3. R. G. Larson, *The Structure and Rheology of Complex Fluids*, (New York: Oxford University Press, 1999).
4. L. D. Landau and I. M. Lifshitz, *Statistical Physics*, (Oxford: Butterworth Heinemann, 1984).
5. T. Kawakatsu, *Statistical Physics of Polymers*, (Berlin: Springer, 2004).

11.4 Exercises

1. Starting with Eq. 11.17 prove Eq. 11.10.

2. Prove that the single chain partition function of a freely jointed polymer chain of size M is

$$q(1, V, T) = V \left(\frac{2\pi l^2}{3} \right)^{3M/2}. \tag{11.47}$$

3. Prove that the free energy of a freely jointed polymer chain of size M is

$$A = -k_B T \ln \left(\frac{q(1, V, T)}{V} P_M(R) \right), \tag{11.48}$$

where $q(1, V, T)$ is the one-chain partition function determined in the previous exercise and $P_M(R)$ is the Gaussian probability distribution defined in Eq. 11.10.

4. Another measure of the size of a polymer chain, besides the root-mean-square end-to-end distance, is the radius of gyration, defined as follows:

$$\langle R_G^2 \rangle = \left\langle \left(\sum_{i=1}^{M-1} |\underline{u}_i - \underline{r}_G|^2 \right) \right\rangle, \tag{11.49}$$

where $\underline{r}_G$ is the center of mass of the polymer.
   Prove that for $M \gg 1$, $\langle R_G^2 \rangle \approx (1/6) M l^2$.

12

Non-equilibrium thermodynamics

In this chapter we examine the non-equilibrium response of a system to external perturbation forces. Example external perturbation forces are electromagnetic fields, and temperature, pressure, or concentration gradients. Under such external forces, systems move away from equilibrium and may reach a new steady state, where the added, perturbing energy is dissipated as heat by the system. Of interest is typically the transition between equilibrium and steady states, the determination of new steady states and the dissipation of energy. Herein, we present an introduction to the requisite statistical mechanical formalism.

12.1 Linear response theory

Let us consider a classical mechanical system of N particles at equilibrium. Let the Hamiltonian of the system be $H_o(\underline{X})$. The system's microscopic states evolve in time according to Hamilton's equations of motion. As we discussed in Chapter 4, an equivalent description is attained by the Liouville equation for the phase space probability distribution, $D(\underline{p}, \underline{q})$, written as

$$\frac{\partial D}{\partial t} + i\mathcal{L}D = 0, \qquad (12.1)$$

where $i\mathcal{L}$ is the Liouville operator.

At equilibrium

$$\frac{\partial D}{\partial t} = 0 \qquad (12.2)$$

and $D_o = D(H_o)$.

Let us assume that the system has been at equilibrium since $t = -\infty$ and let $F(t)$ be a field applied to a system at time $t = 0$. Consider a mechanical property A that is a function of particle momenta and coordinates, $A(\underline{p}, \underline{q})$. Assume that $A(\underline{p}, \underline{q})$ couples and responds to the external perturbation so that the energy of the system is perturbed from its original value as follows:

$$H = H_o - AF(t). \tag{12.3}$$

The average value of the mechanical property A changes from its equilibrium value in response to the external force $F(t)$. We can again consider an ensemble of systems under the influence of the external force and denote the ensemble averaged perturbed value of A as

$$\langle A \rangle = \langle A(t; F) \rangle. \tag{12.4}$$

We limit ourselves to examining the linear response of the system to the external perturbation. This means that $F(t)$ is sufficiently small so that

$$\langle A(t; \lambda F) \rangle = \lambda \langle A(t; F) \rangle, \tag{12.5}$$

where λ is a small positive number (Fig. 12.1).

Property A will change in time until it reaches a new steady state. There are numerous, important questions related to this process: what is this new state? And how can the dynamic response of the system be described in the linear response regime? And if the external force is turned off after the system has been at the steady state for a long time, how does the system return to equilibrium? A central theorem of non-equilibrium statistical mechanics that addresses these questions is

Figure 12.1 Linear response of a macroscopic property A to external perturbations of varying strength.

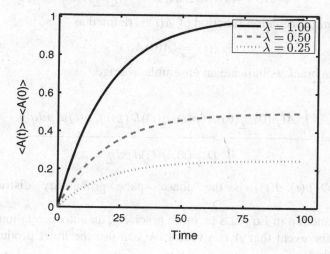

the fluctuation–dissipation theorem. First articulated by Onsager in the 1930s and later refined by Green, Kubo, and Zwanzig, among others, the fluctuation–dissipation theorem states that the response of a system to an external perturbation is related to the internal fluctuations of the system in the absence of the disturbance. The interested reader is referred to the literature at the end of the chapter for numerous informative reviews of the theorem, its development, and its applications.

In what follows, we present elements of linear response theory, and prove the fluctuation–dissipation theorem. We then discuss polymer dielectric spectroscopy as an illustrative example.

First, we need to introduce the concept of time correlation functions in detail.

12.2 Time correlation functions

Let us consider an N-particle system at equilibrium. Let $\underline{p} = \underline{p}(0)$ and $\underline{q} = \underline{q}(0)$ be the values of the $3N$ momenta and coordinates at an initial time $t = 0$.

These change in time according to the equations of motion, so that

$$\underline{p}(t) = \underline{p}(\underline{p}, \underline{q}; t),$$
$$\underline{q}(t) = \underline{q}(\underline{p}, \underline{q}; t). \tag{12.6}$$

Consider a mechanical property A that is a function of the classical degrees of freedom. Then

$$A(t) = A(\underline{p}(t), \underline{q}(t)) = A(\underline{p}, \underline{q}; t). \tag{12.7}$$

The time correlation function of $A(t)$ is defined as

$$C_{AA}(t) = \langle A(0)A(t) \rangle, \tag{12.8}$$

where the brackets indicate an ensemble average.

Then

$$C_{AA}(t) = \frac{\displaystyle\int_{\Gamma} A(\underline{p}(0), \underline{q}(0))A(\underline{p}(t), \underline{q}(t))D(\underline{p}(t), \underline{q}(t))d\underline{p}d\underline{q}}{\displaystyle\int_{\Gamma} D(\underline{p}(t), \underline{q}(t))d\underline{p}d\underline{q}}, \tag{12.9}$$

where $D(\underline{p}(t), \underline{q}(t))$ is the phase space probability distribution function.

The function in Eq. 12.8 is, more precisely, an autocorrelation function. In the event that A is a vector we can use the inner product and write

$$C_{AA}(t) = \langle \underline{A}(0) \cdot \underline{A}(t) \rangle. \tag{12.10}$$

The time correlation function can also be defined between two distinct properties, $A(\underline{p}, \underline{q}; t)$ and $B(\underline{p}, \underline{q}; t)$ as follows:

$$C_{AB}(t) = \langle A(0)B(t) \rangle \tag{12.11}$$

or

$$C_{AB}(t) = \frac{\int_{\Gamma} A(\underline{p}(0), \underline{q}(0))B(\underline{p}(t), \underline{q}(t))D(\underline{p}(t), \underline{q}(t))d\underline{p}d\underline{q}}{\int_{\Gamma} D(\underline{p}(t), \underline{q}(t))d\underline{p}d\underline{q}}. \tag{12.12}$$

The values of A and B fluctuate about their equilibrium values, $\langle A \rangle$ and $\langle B \rangle$, respectively. Let the magnitude of these fluctuations be

$$\delta A(t) = A(t) - \langle A \rangle,$$
$$\delta B(t) = B(t) - \langle B \rangle. \tag{12.13}$$

We have seen in previous chapters that fluctuations are in themselves important. For example, we determined that energy fluctuations in the canonical ensemble are related to the heat capacity of the system.

We can more generally define a non-normalized time correlation function between A at an initial time t' and B at a subsequent time $t' + t$ as follows:

$$C_{AB}(t) = \langle \delta A(t')\delta B(t' + t) \rangle \tag{12.14}$$

or

$$C_{AB}(t) = \langle (A(t') - \langle A \rangle)(B(t' + t) - \langle B \rangle) \rangle. \tag{12.15}$$

At equilibrium the correlation function reduces to

$$C_{AB}(t) = \langle A(t')B(t' + t) \rangle - \langle A \rangle \langle B \rangle. \tag{12.16}$$

Equation 12.11 is then a special case of Eq. 12.16 for $t' = 0$ and $\langle A \rangle = \langle B \rangle = 0$.

It is worthwhile noting here that a logical conjecture of the theory of ergodicity is that equilibrium ensemble averages can be substituted by time averages over all time origins, t'. Hence,

$$C_{AB}(t) = C_{AB}(-t) = \langle \delta A(t')\delta B(t' + t) \rangle = \langle \delta A(t' - t)\delta B(t') \rangle. \tag{12.17}$$

We can write for the autocorrelation function of the fluctuations of A

$$C_{AA}(t) = \langle \delta A(t') \delta A(t' + t) \rangle - \langle A \rangle^2 \qquad (12.18)$$

and

$$C_{AA}(t) = \langle A^2 \rangle - \langle A \rangle^2. \qquad (12.19)$$

We can normalize the autocorrelation functions as follows:

$$c_{AA}(t) = \frac{C_{AA}(t)}{C_{AA}(0)}. \qquad (12.20)$$

Thus,

$$c_{AA}(t) = \frac{\langle \delta A(t') \delta A(t' + t) \rangle}{\langle (\delta A(t'))^2 \rangle}. \qquad (12.21)$$

Since the system is at equilibrium, time is of no importance. Therefore,

$$\langle (\delta A(t'))^2 \rangle = \langle (\delta A)^2 \rangle. \qquad (12.22)$$

In Fig. 12.2, a mechanical property A is shown fluctuating about its equilibrium value. The autocorrelation function of the fluctuations has a value of 1 at the origin,

$$c_{AA}(0) = 1, \qquad (12.23)$$

Figure 12.2 Fluctuations of a mechanical property A at equilibrium.

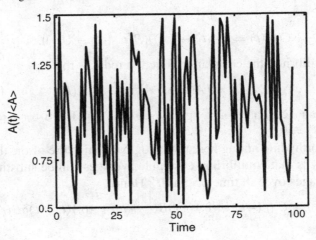

and it decreases at later times, $t > 0$,

$$c_{AA}(t) < 1. \tag{12.24}$$

Onsager surmised that for properties that fluctuate due to thermal motion, the correlation function decays to zero at long time scales,

$$\lim_{t \to \infty} c_{AA}(t) = 0. \tag{12.25}$$

The correlation function provides a measure of how rapidly the values of a mechanical property change as a result of molecular collisions. We can define a characteristic time constant, called the correlation time of property A, that describes the relaxation of $c_{AA}(t)$ from 1 to 0, as follows:

$$\tau_A = \int_0^\infty c_{AA}(t)dt. \tag{12.26}$$

Simple systems with random noise fluctuations may experience exponential decay,

$$c_{AA}(t) \propto \exp(-t/\tau_A). \tag{12.27}$$

In Fig. 12.3, example autocorrelation functions are shown for varying relaxation times. Plotting $\ln c_{AA}(t)$ vs. t would result in a straight line of slope $-1/\tau_A$.

Figure 12.3 Single exponential decay autocorrelation functions.

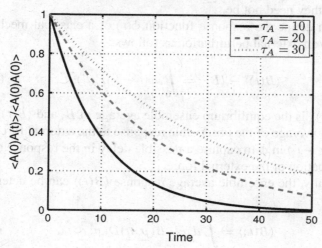

Time correlation functions provide a useful means for elucidating molecular motions and their influence on mechanical properties. They are also useful in the description of numerous interesting phenomena, such as molecular diffusion, electromagnetic radiation, and light scattering.

Importantly, transport coefficients of fluids can be determined with the help of correlation functions. For example, the diffusion coefficient of any simple fluid is related to the particle velocity autocorrelation function as follows:

$$D = \frac{1}{3} \int_0^\infty \langle \underline{v}(0) \cdot \underline{v}(t) \rangle \, dt. \tag{12.28}$$

The interested reader is referred to McQuarrie's book for a detailed exposition of time correlation functions (see Further reading).

12.3 Fluctuation–dissipation theorem

Consider a system at equilibrium with a phase space probability distribution function D_o. If an external field $F(t)$ that couples to a mechanical property $A(p, q)$ is applied at $t = 0$, the system will move away from its equilibrium state. The phase space probability distribution will now become a function of time, $D = D(\underline{p}, \underline{q}; t)$, changing according to Liouville's equation, Eq. 12.1.

Generally, the observable response to an external perturbation is a mechanical property $B(\underline{p}, \underline{q})$. Properties A and B may be identical, although they need not be.

We can define the response function $\phi(t)$ of a classical mechanical system to an external perturbation as follows:

$$\langle B(t) \rangle - \langle B \rangle_o = \int_0^t \phi(t - t') F(t') dt', \tag{12.29}$$

where $\langle B \rangle_o$ is the equilibrium ensemble average of B, and $\langle B(t) \rangle$ is the ensemble average of the observable response to the external force field. We use $(t - t')$ in ϕ to capture a possible delay in the response of B to the application of the external force.

Generally, the ensemble average response $\langle B(t) \rangle$ can be determined with

$$\langle B(t) \rangle = \int_\Gamma d\underline{p} d\underline{q} \, B(\underline{p}, \underline{q}) D(\underline{p}, \underline{q}; t). \tag{12.30}$$

In the linear response limit of a small external field, we may assume that the probability distribution function is determined as

$$D = D_o + \Delta D, \tag{12.31}$$

where ΔD is a small perturbation in the probability function.

Using Eqs. 12.1, 12.3, and 12.29, we can determine the small perturbation ΔD as a function of time to terms in first order of the perturbation as follows:

$$\frac{\partial \Delta D}{\partial t} = \{H_o, \Delta D\} - F(t)\{A, D_o\}, \tag{12.32}$$

where the brackets are the Poisson brackets introduced in Chapter 4.

In terms of the Liouville operator, Eq. 12.32 is written as

$$\frac{\partial \Delta D}{\partial t} = -i\mathcal{L}_o \Delta D - F(t)\{A, D_o\}, \tag{12.33}$$

where $i\mathcal{L}_o$ is Liouville's operator at equilibrium.

Integrating Eq. 12.33 in time yields

$$\Delta D(t) = -\int_0^t e^{i(t-t')\mathcal{L}_o}\{A, D_o\}F(t')dt'. \tag{12.34}$$

Assuming that at equilibrium $\langle B \rangle_o = 0$ and combining Eqs. 12.30, 12.31, and 12.34 yields

$$\langle B(t) \rangle = -\int_0^t \int_\Gamma d\underline{p}d\underline{q}B(\underline{p}, \underline{q})e^{i(t-t')\mathcal{L}_o}\{A, D_o\}F(t'). \tag{12.35}$$

Finally, for a canonical ensemble distribution

$$\{A, D_o\} = \beta \dot{A}D_o. \tag{12.36}$$

Combining Eqs. 12.29, 12.35, and 12.36 yields for the response function

$$\phi(t) = \beta \int_\Gamma d\underline{p}d\underline{q}D_o(\dot{A}(0)B(t)), \tag{12.37}$$

or, more simply

$$\phi(t) = \beta \langle \dot{A}(0)B(t) \rangle. \tag{12.38}$$

This is a remarkable equation, known as the fluctuation–dissipation theorem. According to this theorem, the linear response of a system to a small external perturbation is directly related to internal fluctuations of the system at equilibrium.

Another form of the fluctuation–dissipation theorem involves time correlation functions. If the external perturbation is applied at $t = -\infty$ and turned off at $t = 0$, the system will move from a steady state back to equilibrium. The dynamic relaxation of property A from its original steady state, $\langle A(0) \rangle$, to equilibrium, $\langle A \rangle_o$, can be shown to be congruent to the normalized time autocorrelation function of equilibrium fluctuations,

$$\frac{\langle A(t) \rangle - \langle A \rangle_o}{\langle A(0) \rangle - \langle A \rangle_o} = \frac{C_{AA}(t)}{C_{AA}(0)}. \tag{12.39}$$

In the following section we illustrate these concepts with the example of polymer dielectric relaxation.

12.4 Dielectric relaxation of polymer chains

Dielectric spectroscopy is a valuable tool for studying the conformational and dynamic properties of polar macromolecules. The conformational features can be determined by dielectric relaxation strength measurements, whereas the dielectric spectrum provides information on the dynamics of the macromolecules. Phenomenological and molecular theories of dielectric permittivity and dielectric relaxation of polymers have also been developed to elucidate the experimentally observed phenomena. As Adachi and Kotaka have stressed (see Further reading), experimental information depends on each monomer's dipole vector direction as related to the chain contour. A classification of polar polymers into three categories was introduced by Stockmayer: type-A polymers, where the dipole is parallel to the chain contour (Fig. 12.4), type-B, where it is perpendicular to the chain contour, and type-C, where the dipoles are located on mobile side groups. For type-A chains, the global dipole moment of each chain is directly proportional to the chain's end-to-end vector $\underline{R}$.

In the presence of an external electric field, polar chains reorient themselves in the direction of the field (Fig. 12.5). The orientational motions of polar macromolecules give rise to dielectric absorption. The rotational rearrangement of macromolecules in a solution takes a finite time, increasing the polarization $\underline{P}$ of the system in the direction of the field, where $\underline{P}$ is the sum of the constituent dipoles per unit volume V:

$$\underline{P} = \frac{\sum\limits_{i=1}^{M_c} \underline{M}_i}{V},$$ (12.40)

where $\underline{M}_i$ is the total dipole of the ith chain and M_c is the number of chains in the system. This increase of polarization with time after application of a field is called the build-up process, and is described by the normalized build-up function $\phi_P(t)$. When the electric field is turned off, Brownian motion randomizes the orientation of the chains to the equilibrium distribution. This decay process is described by the

Figure 12.4 Type-A polar polymer.

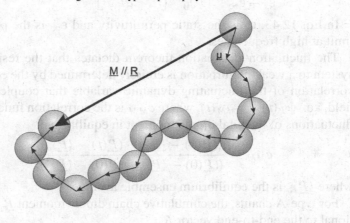

Figure 12.5 Polar polymer reorientation due to applied electric field.

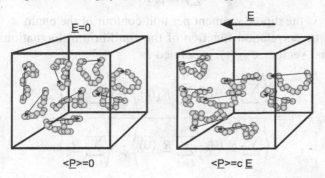

normalized decay function $\psi_P(t)$, where

$$\psi_P(t) = 1 - \phi_P(t) = \frac{|\Delta \underline{P}(t)|}{|\Delta \underline{P}(0)|}, \tag{12.41}$$

with $\Delta \underline{P}(t) = \langle \underline{P}(t) \rangle - \langle \underline{P} \rangle_o$ being the departure of the polarization from its equilibrium value.

The experimentally observed quantity is the complex dielectric permittivity, defined as

$$\varepsilon^*(\omega) = \varepsilon'(\omega) - i\varepsilon''(\omega), \tag{12.42}$$

with ε' the real permittivity, ε'' the dielectric loss, and ω the characteristic frequency. Note that $\varepsilon^*(\omega)$ is given by the Laplace–Fourier transform of the normalized decay function $\psi_P(t)$

$$\frac{\varepsilon^*(\omega) - \varepsilon_\infty}{\varepsilon_s - \varepsilon_\infty} = \int_0^\infty \left[-\frac{d\psi_P(t)}{dt} \right] \exp(-i\omega t)dt. \tag{12.43}$$

In Eq. 12.43, ε_s is the static permittivity and ε_∞ is the permittivity limit at high frequencies.

The fluctuation–dissipation theorem dictates that the response of a system to a weak perturbation is entirely determined by the equilibrium correlation of the fluctuating dynamic variable that couples with the field, i.e., $\psi_P(t) = c_{MM}(t)$, where c_{MM} is the correlation function of the fluctuations of the total dipole moment in equilibrium:

$$c_{MM} = \frac{\langle (\underline{P}(0) - \langle \underline{P} \rangle_o) \cdot (\underline{P}(t) - \langle \underline{P} \rangle_o) \rangle}{\langle (\underline{P}(0) - \langle \underline{P} \rangle_o) \cdot (\underline{P}(0) - \langle \underline{P} \rangle_o) \rangle}, \tag{12.44}$$

where $\langle \underline{P} \rangle_o$ is the equilibrium ensemble average.

For type-A chains, the cumulative chain dipole moment $\underline{P}_i$ is proportional to the end-to-end vector $\underline{R}_i$,

$$\underline{P}_i = \mu' \underline{R}_i, \tag{12.45}$$

where μ' is the dipole moment per unit contour of the chain.

The time correlation function of the equilibrium fluctuations of the end-to-end vectors, $c_{RR}(t)$, is defined as

$$c_{RR}(t) = \frac{\left\langle \sum_{i=1}^{M_c} \underline{R}_i(0) \cdot \sum_{k=1}^{M_c} \underline{R}_k(t) \right\rangle - \left\langle \sum_{i=1}^{M_c} \underline{R}_i \right\rangle^2}{\left\langle \sum_{i=1}^{M_c} \underline{R}_i(0) \cdot \sum_{k=1}^{M_c} \underline{R}_k(0) \right\rangle - \left\langle \sum_{i=1}^{M_c} \underline{R}_i \right\rangle^2}, \tag{12.46}$$

and it is equal to c_{MM}.

Consequently, according to the fluctuation–dissipation theorem, one has

$$\psi_P(t) = c_{RR}(t). \qquad (12.47)$$

As a result, the dielectric relaxation reflects the fluctuations of $\underline{R}$, establishing a connection between dielectric absorption and the equilibrium reorientational motion of polymers.

The complex dielectric permittivity can then be calculated as the Laplace–Fourier transform of c_{RR} and conversely macromolecular dynamic features can be revealed by measuring $\varepsilon^*(\omega)$. This indeed is a main focus of dielectric spectroscopy experiments.

For systems in homogeneous environments and in the absence of cross-correlations between different chains, c_{RR} is further reduced to the autocorrelation function of the end-to-end vector,

$$c_{RR}(t) = \frac{\langle \underline{R}(0) \cdot \underline{R}(t) \rangle}{\langle \underline{R}(0) \cdot \underline{R}(0) \rangle}. \qquad (12.48)$$

An illustration of the congruency between the linear response of type-A polymer chains to external electric fields and the equilibrium autocorrelation function of the end-to-end vector was provided in the work of the author with Ed Maginn and Davide Hill (see references in Further reading).

12.5 Further reading

1. I. Prigogine, *Non-Equilibrium Statistical Mechanics*, (New York: Interscience Publishers, 1962).
2. J. McConnell, *Rotational Brownian Motion and Dielectric Theory*, (New York: Academic Press, 1980).
3. D. A. McQuarrie, *Statistical Mechanics*, (Sausalito, CA: University Science Books, 2000).
4. L. Onsager, *Phys. Rev.*, **37**, 405, (1931).
5. L. Onsager, *Phys. Rev.*, **38**, 2265, (1931).
6. R. Kubo, *Rep. Prog. Phys.*, **29**, 255, (1966).
7. Y. N. Kaznessis, D. A. Hill, and E. J. Maginn, *J. Chem. Phys.*, **109**, 5078, (1998).
8. Y. N. Kaznessis, D. A. Hill, and E. J. Maginn, *Macromolecules*, **31**, 3116, (1998).
9. Y. N. Kaznessis, D. A. Hill, and E. J. Maginn, *Macromolecules*, **32**, 1284, (1999).
10. K. Adachi and T. Kotaka, *Macromolecules*, **21**, 157, (1988).
11. K. Adachi and T. Kotaka, *Macromolecules*, **16**, 1936, (1983).

12.6 Exercises

1. Generate a randomly changing variable and calculate the autocorrelation function of the fluctuations. Investigate the influence of the fluctuations' magnitude and frequency on the relaxation time.

2. Prove Eq. 12.16.

3. Prove Eq. 12.17.

4. Prove Eq. 12.32.

5. Prove Eq. 12.36.

13

Stochastic processes

There are important engineering and physical processes that appear random. A canonical example is the motion of Brownian particles, discussed in Chapter 11. In random or stochastic processes, in contrast to deterministic ones, there is not one single outcome in the time evolution of a system, even when initial conditions remain identical. Instead there may be different outcomes, each with a certain probability. In this chapter we present stochastic processes and derive a general framework for determining the probability of outcomes as a function of time.

We are starting the discussion with reacting systems away from the thermodynamic limit. Typically, reacting systems are modeled with ordinary differential equations that express the change of concentrations in time as a function of reaction rates. This continuous and deterministic modeling formalism is valid at the thermodynamic limit. Only when the number of molecules of reacting species is large enough can the concentration be considered a continuously changing variable. Importantly, the reaction events are considered to occur deterministically at the thermodynamic limit. This means that there is certainty about the number of reaction events per unit time and unit volume in the system, given the concentration of reactants.

On the other hand, if the numbers of reacting molecules are very small, for example in the order of $O(10^{-22}N_A)$, then integer numbers of molecules must be modeled along with discrete changes upon reaction. Importantly, the reaction occurrences can no longer be considered deterministic, but probabilistic. In this chapter we present the theory to treat reacting systems away from the thermodynamic limit. We first present a brief overview of continuous-deterministic chemical kinetics models and then discuss the development of stochastic-discrete models.

We then generalize the discussion to other stochastic processes, deriving appropriate modeling formalisms such as the master equation, the Fokker–Planck equation and the Langevin equation.

13.1 Continuous-deterministic reaction kinetics

Consider the simple reversible reaction between two chemical compounds A and B,

$$A \underset{k_{-1}}{\overset{k_1}{\rightleftharpoons}} B, \qquad (13.1)$$

where k_1, k_{-1} are the forward and reverse reaction rate constants, respectively. Assume that the reacting system is at constant temperature T in constant volume V.

The reaction rates $-r_A$ and $-r_B$ can be defined as the change in the number of moles of A and B, respectively, per unit volume per unit time. By definition

$$-r_A = -k_1 C_A + k_{-1} C_B \qquad (13.2)$$

and

$$-r_B = k_1 C_A - k_{-1} C_B. \qquad (13.3)$$

In Eqs. 13.2 and 13.3, C_i is the concentration of species i, where $i = A, B$, in the system volume V. These are defined as follows

$$C_i = \frac{N_i}{N_{\text{Avogadro}} V}, \qquad (13.4)$$

where N_i, $i = A, B$ is the number of molecules of species i in V.

The mole balances on species A and B are written as

$$\frac{dC_A}{dt} = -r_A$$
$$= -k_1 C_A + k_{-1} C_B, \qquad (13.5)$$

and

$$\frac{dC_B}{dt} = -r_B$$
$$= k_1 C_A - k_{-1} C_B, \qquad (13.6)$$

respectively.

In this simple case of a reversible reaction, only one of the two concentrations is an independent variable, say C_A, with C_B completely determined by C_A and the initial conditions.

The ordinary differential equation, Eq. 13.5, can be written in terms of numbers of molecules, multiplying both sides by the constant volume,

V, to yield

$$\frac{dN_A}{dt} = -k_1 N_A + k_{-1} N_B. \tag{13.7}$$

Equation 13.7 can be solved with initial conditions $N_{io} = N_i(t = 0)$.
It is convenient to introduce a new variable, the conversion of A, defined as

$$\Psi = \frac{N_{Ao} - N_A}{N_{Ao}}. \tag{13.8}$$

The following stoichiometric table summarizes the changes upon reaction in terms of Ψ.

Species	Initial	Change	Remaining
A	N_{Ao}	$-N_{Ao}\Psi$	$N_A = N_{Ao}(1 - \Psi)$
B	N_{Bo}	$N_{Ao}\Psi$	$N_B = N_{Bo} + N_{Ao}\Psi$
Total	$N = N_{Ao} + N_{Bo}$		N

Equation 13.7 can then be written as

$$-N_{Ao}\frac{d\Psi}{dt} = -k_1 N_{Ao}(1 - \Psi) + k_{-1}(N_{Bo} + N_{Ao}\Psi). \tag{13.9}$$

With a trivial initial condition, $\Psi(0) = 0$, the solution for Ψ is

$$\Psi = \frac{e^{-(k_1 + k_{-1})t} + \left(k_1 - \frac{N_{Bo}}{N_{Ao}}k_{-1}\right)}{(k_1 + k_{-1})}. \tag{13.10}$$

The number of molecules of A change in time according to

$$N_A = N_{Ao}(1 - \Psi). \tag{13.11}$$

At equilibrium, $dN_A/dt = dN_B/dt = 0$, and the chemical potential of reactants is the same, $\mu_A = \mu_B$. We write

$$\frac{N_B}{N_A} = \frac{k_1}{k_{-1}} = K_{eq}, \tag{13.12}$$

where K_{eq} is the equilibrium constant.

This discussion can be naturally extended to multicomponent systems involving multiple reactions. Importantly, as long as the number of molecules of reactants and products is large, reactions are regarded as deterministic rate processes and reaction rate equations are written for each chemical species with ordinary differential equations, similar to Eqs. 13.5 and 13.6.

13.2 Away from the thermodynamic limit – chemical master equation

The assumption underlying traditional chemical reaction kinetics, as illustrated in the preceding section, is that the system is at the thermo-dynamic limit, i.e., $N_i \to \infty$ for all reacting species in volume V with $V \to \infty$. This assumption has the following two important implications:

1. Although the number of molecules N_i is a discrete variable, we can assume it to be changing continuously. It is indeed impossible to distinguish between two systems that have for example 10^{23} and $10^{23} + 1$ particles.
2. Although the reactions are discrete events, they are regarded as continuous-deterministic rate processes.

At the thermodynamic limit then, the continuous-deterministic model in Eq. 13.7 can validly capture the reaction dynamics.

Figure 13.1 Reacting systems at and away from the thermodynamic limit.

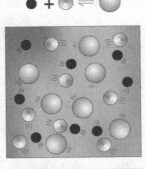

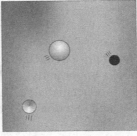

On the other hand, if the system is far from the thermodynamic limit with the number N_i of chemical species being relatively small, the system behaves probabilistically and discretely (Fig. 13.1). In the 1940s, McQuarrie, among others, developed a general theory for such stochastic-discrete reaction kinetics. In what follows we present the essential elements of this theory. The interested reader is referred to the literature in Further reading for a more detailed exposition of the subject.

Suppose a system of N species $(S_1, \ldots, S_N)$ is reacting through M reactions $(R_1, \ldots, R_M)$ in volume V. Define the state of the system at any time t with the N-dimensional vector

$$\underline{X}(t) = [X_1(t), \ldots, X_N(t)], \qquad (13.13)$$

where $X_i(t)$ is the number of molecules of species S_i at time t. The state space of the system is then N-dimensional, positive, and discrete (Fig. 13.2).

Define an $M \times N$ stoichiometric matrix $\underline{\underline{\nu}}$, where ν_{ij} is the stoichiometric coefficient of species S_i in reaction $\overline{\overline{R}}_j$.

Define $a_j(\underline{X})dt$ as the probability, given the system is at $\underline{X}(t)$ at time t, that a reaction R_j will occur somewhere in the system volume in the time interval $[t, t + dt]$. The propensity, a_j, is defined as $a_j = h_j k_j$, where h_j is the number of possible combinations of the reacting molecules involved in R_j, and k_j is the reaction rate constant. This will likely become more clear with an example, presented in the next section.

Figure 13.2 Discrete state space.

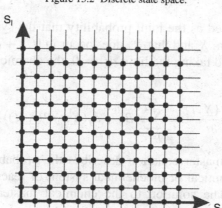

Transitions from one state to another occur when a reaction event takes place, consuming the reactant molecules and creating the product molecules. We shall see that all transitions have a propensity that is analogous to a mesoscopic reaction rate.

All of the possible states of the system form the state space of the system. The current position of the system in the state space is called the state point. In a stochastic system, each point in state space, $\underline{X}$, will have a certain probability at any particular moment in time. Let us denote this probability at time t with $P(\underline{X}; t)$. Of course, the sum of the probabilities of all points in state space at the same time t is equal to one.

One can derive an equation governing how the probability distribution of the state space changes over time, called the master equation, by applying the laws of conservation of probability to each possible state of the system.

The probability of the system being in state $\underline{X}$ at time $t + dt$ is $P(\underline{X}; t + dt)$. We can define the interval dt to be sufficiently small so that at most one reaction can occur during dt. The probability is then determined as the sum of two terms: the first term is the probability of the system being in state $\underline{X}$ at time t and no reaction occurring in the interval $[t, t + dt]$; the second term is the probability of the system being one reaction away from $\underline{X}$ and that reaction occurring in the interval $[t, t + dt]$.

We can write

$$P(\underline{X}; t + dt) = P(\underline{X}; t) \left(1 - \sum_{j=1}^{M} a_j(\underline{X}; t)dt \right) + \sum_{j=1}^{M} B_j dt, \quad (13.14)$$

where B_j is defined as the joint probability that the system is one R_j reaction away from $\underline{X}$ and that R_j does occur in $[t, t + dt]$.

Rearranging and taking the limit $dt \to 0$, the chemical master equation is obtained,

$$\frac{\partial P(\underline{X}, t)}{\partial t} = \sum_{j=1}^{M} \left(B_j - a_j(\underline{X}; t)P(\underline{X}, t) \right). \quad (13.15)$$

The chemical master equation describes the probabilistic dynamic behavior of a chemical or biochemical system of reactions in a well-mixed volume. The solution to the chemical master equation is a probability distribution of finding the system within the state space at a specific time.

Analytical solutions to the chemical master equation are only available for extremely simple systems. Furthermore, numerical solutions to the master equation are computationally intractable for all but the simplest systems, because as the number of unique chemical species in the system grows linearly, the number of possible states in the phase space grows combinatorially. Enumerating the transitions between states then becomes difficult.

In the next section, we present a simple example that accepts an analytic solution of the master equation, in order better to illustrate the derivation and solution of Eq. 13.15.

13.2.1 Analytic solution of the chemical master equation

Consider a spatially homogeneous system of volume V. Suppose the simple reaction R_1 is taking place in the system, where species S_1 decomposes irreversibly to species S_2 and S_3,

$$S_1 \xrightarrow{k_1} S_2 + S_3. \tag{13.16}$$

Define a constant c_1 so that

$$
\begin{aligned}
c_1 dt = \ & \text{average probability that a particular molecule} \\
& \text{of species 1 will react according to } R_1 \\
& \text{in the next infinitesimal time interval } dt. \quad (13.17)
\end{aligned}
$$

If there are X_1 molecules of species S_1 in volume V at time t, define

$$
\begin{aligned}
X_1 c_1 dt = \ & \text{probability that any molecule} \\
& \text{of species 1 will react according to } R_1 \\
& \text{in the next infinitesimal time interval } dt. \quad (13.18)
\end{aligned}
$$

It turns out that constant c_1 is equal to the kinetic reaction constant k_1 in this simple case. Let us prove this.

The average rate at which R_1 occurs in V is

$$\langle X_1 c_1 \rangle = c_1 \langle X_1 \rangle. \tag{13.19}$$

The average reaction rate per unit volume is

$$r_1 = \frac{c_1 \langle X_1 \rangle}{V} \tag{13.20}$$

or, using the concentration $[X_1]$ of species S_1,

$$r_1 = c_1[X_1]. \tag{13.21}$$

By definition, the kinetic constant k_1 is the average reaction rate per unit volume divided by the concentration of S_1,

$$k_1 = \frac{r_1}{[X_1]}. \tag{13.22}$$

Clearly then $c_1 = k_1$.

Generally, for any reaction R_μ we can define a constant c_μ so that

$c_\mu dt$ = average probability that a particular combination

  of reactant molecules will react according to R_μ

  in the next infinitesimal time interval dt. $\qquad$ (13.23)

With simple arguments we can show that c_μ is always related to the kinetic constant k_μ of elementary reaction R_μ up to constant factors. For example, for a bimolecular reaction

$$S_1 + S_2 \xrightarrow{k_1} S_3, \tag{13.24}$$

we find $c_1 = k_1/V$, whereas for the reaction

$$2S_1 \xrightarrow{k_1} S_2 + S_3, \tag{13.25}$$

we find $c_1 = 2k_1/V$.

Returning to the example reaction

$$S_1 \xrightarrow{k_1} S_2 + S_3, \tag{13.26}$$

the chemical master equation derivation starts with the following simple equation:

$$P(X;t+dt) = P(X;t)(1 - a_1 dt) + B_1 dt. \tag{13.27}$$

The propensity a_1 is defined so that $a_1 dt$ is the probability that a reaction R_1 will occur in $[t, t+dt]$ given the system is at state X at time t. Then, clearly

$$a_1 dt = X c_1 dt, \tag{13.28}$$

where, in this case, $c_1 = k_1$.

The joint probability $B_1 dt$ that the system is one R_1 away from X at time t and the reaction occurs in $[t, t+dt]$ is given by

$$B_1 dt = P(X+1;t)(X+1)c_1 dt. \tag{13.29}$$

Consequently,

$$P(X; t + dt) = P(X; t) - P(X; t)Xc_1 dt + P(X + 1; t)(X + 1)c_1 dt.$$

(13.30)

Rearranging and taking the limit $dt \to 0$ yields the master equation that describes the change of the probability of the system being anywhere in the available state space as a function of time,

$$\frac{\partial P(X; t)}{\partial t} = c_1 \left[(X + 1)P(X + 1; t) - XP(X; t) \right].$$

(13.31)

Assuming that the initial state of the system is precisely known, we can write for the initial condition

$$P(X; 0) = \delta(X, X_o).$$

(13.32)

This means that there are exactly X_o molecules of S_1 at $t = 0$. Generally, the initial condition may be any probability distribution.

Solving Eq. 13.31 is a taxing task, requiring the definition of probability generating functions. We illustrate the use of generating functions in the solution of this simple example. A detailed exposition of generating functions and their use in solving master probability equations is beyond the scope of this text. The interested reader is referred to the book by Daniel Gillespie (see Further reading).

If X is a random variable in time with probability $P(X; t)$, the generating function of X is defined as a power series

$$G(s; t) = \sum_{X=0}^{\infty} s^X P(X; t).$$

(13.33)

From the definition, the next few relations follow:

$$\frac{\partial G(s; t)}{\partial s} = \sum_{X=0}^{\infty} s^X (X + 1)P(X + 1; t),$$

(13.34)

$$sG(s; t) = \sum_{X=0}^{\infty} s^X P(X - 1; t),$$

(13.35)

$$s\frac{\partial G(s; t)}{\partial s} = \sum_{X=0}^{\infty} s^X XP(X + 1; t),$$

(13.36)

$$s^2 \frac{\partial G(s; t)}{\partial s} = \sum_{X=0}^{\infty} s^X (X - 1)P(X - 1; t).$$

(13.37)

We can now start with Eq. 13.31 and write

$$\sum_{X=0}^{\infty} s^X \frac{\partial P(X;t)}{\partial t} = \sum_{X=0}^{\infty} s^X c_1(X+1)P(X+1;t) - \sum_{X=0}^{\infty} s^X c_1 X P(X;t).$$

(13.38)

Using generating function relations yields

$$\frac{\partial G(s;t)}{\partial t} = c_1(1-s)\frac{\partial G(s;t)}{\partial s},$$

(13.39)

with $G(s;0) = s^{X_o}$ for all s.

The chemical master equation solution is found by first solving the partial differential equation 13.39 and then transforming back to probability space. Finally,

$$P(X;t) = \frac{X_o!}{X!(X_o - X)!}e^{-c_1 X t}\left(1 - e^{-c_1 t}\right)^{X_o - X}.$$

(13.40)

In Fig. 13.3, the probability distribution is shown as a function of time and of the number of molecules for a simple example with $X_o = 10$ and $c_1 = 0.1$.

The average number of molecules X is

$$\langle X(t)\rangle = \int_0^{\infty} X P(X;t)dX,$$

(13.41)

Figure 13.3 The probability as a function of time of the number of molecules X of S_1 involved in reaction $S_1 \xrightarrow{k_1} S_2 + S_3$. In this example, $X_o = 10$ and $c_1 = 0.1$ in reduced units. In the insert the deterministic solution is shown.

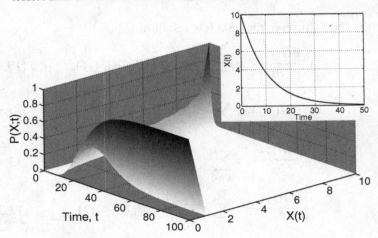

which yields

$$\langle X \rangle = X_o e^{-c_1 t}. \tag{13.42}$$

Similarly, the standard deviation is determined as

$$\Delta(t) = \left(X_o e^{-c_1 t} (1 - e^{-c_1 t}) \right)^{1/2}. \tag{13.43}$$

In this trivial example, $c_1 = k_1$. It is clear that the average X is equal to the solution of the continuous-deterministic reaction rate law equation

$$\frac{d[X]}{dt} = -k_1 X. \tag{13.44}$$

Of course, the continuous-deterministic solution does not afford for calculation of variances and fluctuations around the mean. Furthermore, for non-linear systems the solution of the ordinary differential solution may not even accurately capture the average concentration, resulting in incorrect results.

Unfortunately, there is no tractable solution of the master equation for more complex systems of higher order reactions. In such cases, one can resort to numerical simulations that sample the probability distribution. We discuss these in Chapter 18.

13.3 Derivation of the master equation for any stochastic process

We can now look beyond chemical reacting systems and consider any arbitrary stochastic process $\underline{X}(t)$, indexed by time t, as the sequence of random variables, $X(t_1), X(t_2), \ldots, X(t_n)$, observed at times $t_1 < t_2 < \cdots < t_n$.

Define the joint probability that the variable attains specific values $X(t_1) = x_1; X(t_2) = x_2, \ldots, X(t_n) = x_n$ as

$$P_n(x_1, t_1; x_2, t_2; \ldots; x_n, t_n). \tag{13.45}$$

The conditional probability that $X(t_2) = x_2$, provided that $X(t_1) = x_1$, is

$$P(x_2, t_2 | x_1, t_1) = \frac{P_2(x_1, t_1; x_2, t_2)}{P_1(x_1, t_1)}. \tag{13.46}$$

Generally, the conditional probability that $X(t_{k+1}) = x_{k+1}; X(t_{k+2}) = x_{k+2}, \ldots, X(t_{k+l}) = x_{k+l}$ provided that $X(t_1) = x_1; X(t_2) = x_2; \ldots;$

$X(t_k) = x_k$ is

$$P(x_{k+1}, t_{k+1}; x_{k+2}, t_{k+2}; \dots; x_{k+l}, t_{k+l} | x_1, t_1; x_2, t_2; \dots; x_k, t_k)$$
$$= \frac{P_{k+l}(x_1, t_1; x_2, t_2; \dots; x_{k+l}, t_{k+l})}{P_k(x_1, t_1; x_2, t_2; \dots; x_k, t_k)}. \tag{13.47}$$

A stochastic process is called a Markov process if for any set of n successive times we have

$$P(x_n, t_n | x_1, t_1; x_2, t_2; \dots; x_{n-1}, t_{n-1}) = P(x_n, t_n | x_{n-1}, t_{n-1}). \tag{13.48}$$

This means that for a Markov stochastic process the probability of each outcome depends only on the immediately previous outcome.

A Markov process is completely determined if $P_1(x_1, t_1)$ and the one-step conditional, or transition, probability $P(x_{k+1}, t_{k+1} | x_k, t_k)$ are known. For instance, then

$$P_3(x_1, t_1; x_2, t_2; x_3, t_3) = P_2(x_1, t_1; x_2, t_2)P(x_3, t_3 | x_2, t_2) \tag{13.49}$$

and

$$P_3(x_1, t_1; x_2, t_2; x_3, t_3) = P_1(x_1, t_1)P(x_2, t_2 | x_1, t_1)P(x_3, t_3 | x_2, t_2). \tag{13.50}$$

All joint probabilities can be determined similarly for an arbitrary number of time points, n.

13.3.1 Chapman–Kolmogorov equation

Let us integrate Eq. 13.49 over all possible x_2 to obtain

$$P_2(x_1, t_1; x_3, t_3) = P_1(x_1, t_1) \int P(x_2, t_2 | x_1, t_1)P(x_3, t_3 | x_2, t_2)dx_2. \tag{13.51}$$

Dividing by $P_1(x_1, t_1)$ we obtain the Chapman–Kolmogorov equation, which is obeyed by any Markov process,

$$P(x_3, t_3 | x_1, t_1) = \int P(x_2, t_2 | x_1, t_1)P(x_3, t_3 | x_2, t_2)dx_2. \tag{13.52}$$

The Chapman–Kolmogorov equation makes obvious sense: a process starting with a value x_1 at time t_1 can reach a value x_3 at a later time t_3 through any one of the possible values at intermediate times, t_2 (Fig. 13.4).

The following identities are true:

$$\int P(x_2, t_2|x_1, t_1)dx_2 = 1 \qquad (13.53)$$

and

$$\int P(x_2, t_2|x_1, t_1)P_1(x_1, t_1)dx_2 = P_1(x_2, t_2). \qquad (13.54)$$

If the single-step transition probability only depends on the time interval $t = t_2 - t_1$, we can write more simply

$$P(x_2, t_2|x_1, t_1) = T_t(x_2|x_1). \qquad (13.55)$$

The Chapman–Kolmogorov equation can then be written in terms of transition probabilities as follows:

$$T_{t+t'}(x_3|x_1) = \int T_{t'}(x_3|x_2)T_t(x_2|x_1)dx_2. \qquad (13.56)$$

13.3.2 Master equation

We define the transition probability per unit time $W(x_3|x_2)$ as follows:

$$W(x_3|x_2) = \left.\frac{\partial T_{t'}(x_3|x_2)}{\partial t'}\right|_{t'=0}. \qquad (13.57)$$

We can then write

$$T_{t'}(x_3|x_2) = t'W(x_3|x_2) + (1 - a_o(x_3)t')\delta(x_3 - x_2). \qquad (13.58)$$

In Eq. 13.58, $(1 - a_o t')$ is the probability that no transition occurs during the interval t'. The transition propensity a_0 is defined as

$$a_o(x) = \int W(x'|x)dx', \qquad (13.59)$$

Figure 13.4 Chapman–Kolmogorov transitions.

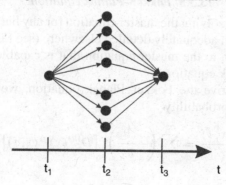

so that $a_o(x)t'$ is the probability of the system being at point x and transitioning away from it during the next interval t'.

Starting with Eq. 13.56, we can now write

$$T_{t+t'}(x_3|x_1) = \int [(1 - a_o(x_3)t')\delta(x_3 - x_2)T_t(x_2|x_1)dx_2$$
$$+ t'W(x_3|x_2)T_t(x_2|x_1)]dx_2 \qquad (13.60)$$

and

$$T_{t+t'}(x_3|x_1) = (1 - a_o(x_3)t')T_t(x_3|x_1)dx_2 + t' \int W(x_3|x_2)T_t(x_2|x_1)dx_2. \qquad (13.61)$$

Dividing by t' and taking the limit as $t' \to 0$,

$$\frac{\partial T_t(x_3|x_1)}{\partial t} = -a_o(x_3)T_t(x_3|x_1)dx_2 + \int W(x_3|x_2)T_t(x_2|x_1)dx_2. \qquad (13.62)$$

Using Eq. 13.59 we write the differential form of the Chapman–Kolmogorov equation, or the master equation of the stochastic process $X(t)$, as follows:

$$\frac{\partial T_t(x_3|x_1)}{\partial t} = \int [W(x_3|x_2)T_t(x_2|x_1) - W(x_2|x_3)T_t(x_3|x_1)] \, dx_2. \qquad (13.63)$$

Further simplification, introducing back the probability, yields the more frequently used form of the master equation

$$\frac{\partial P(x, t)}{\partial t} = \int [W(x|x')P(x', t) - W(x'|x)P(x, t)] \, dx'. \qquad (13.64)$$

The master equation describes the time evolution of the probability in terms of the transition rates of the Markov process.

13.3.3 Fokker–Planck equation

The difficulties in solving the master equation for any but the simplest of systems have been adequately detailed elsewhere (see Further reading). An approximation to the master equation that is capable of solution is the Fokker–Planck equation.

In order to derive the Fokker–Planck equation, we may write an expansion in the probability

$$\frac{\partial P(x, t)}{\partial t} = \sum_{n=1}^{\infty} \left(-\frac{\partial}{\partial x}\right)^n [D^{(n)}(x)P(x, t)], \qquad (13.65)$$

where

$$D^{(n)}(x) = \frac{1}{n!} \lim_{\Delta t \to 0} \frac{1}{\Delta t} < [x(t + \Delta t) - x(t)]^n >. \quad (13.66)$$

The expansion in Eq. 13.65 is known as the Kramers–Moyal expansion. Assuming that $D^{(n)}(x) = 0$ for $n > 2$, yields the Fokker–Planck equation

$$\frac{\partial P(x, t)}{\partial t} = -\frac{\partial}{\partial x}[A(x)P(x, t)] + \frac{\partial^2}{\partial x^2}[B(x)P(x, t)], \quad (13.67)$$

where

$$A(x) = D^{(1)}(x) = \frac{\partial < x >}{\partial t} \quad (13.68)$$

is called the drift coefficient, and

$$B(x) = D^{(2)}(x) = \frac{1}{2}\frac{\partial \sigma^2}{\partial t} \quad (13.69)$$

is called the diffusion coefficient, and $< x >$ and σ^2 are the average and variance of x, respectively.

The Fokker–Planck equation accurately captures the time evolution of stochastic processes whose probability distribution can be completely determined by its average and variance. For example, stochastic processes with Gaussian probability distributions, such as the random walk, can be completely described with a Fokker–Planck equation.

13.3.4 Langevin equation

An equivalent approach to the Fokker–Planck equation for determining the time evolution of a stochastic process is based on the Langevin equation.

In Chapter 11, we discussed how Paul Langevin devised an equation that describes the motion of a Brownian particle (Eq. 11.33). Similar equations can be devised to describe the time evolution of any stochastic process $X(t)$.

The general Langevin equation has the form

$$\frac{dX}{dt} = f(X, t) + g(X, t)\Gamma(t). \quad (13.70)$$

Function $f(X, t)$ is called the drift term and function $g(X, t)$ is called the diffusion term. The Langevin force $\Gamma(t)$ is generally assumed to be a Gaussian random variable with zero mean and δ correlation:

$$\langle \Gamma(t) \rangle = 0 \tag{13.71}$$

and

$$\langle \Gamma(t)\Gamma(t') \rangle = 2\gamma\delta(t - t'), \tag{13.72}$$

where γ is the noise strength, which without loss of generality we can consider equal to one.

Practically, Γ is defined as a normal random variable with zero mean and variance of $1/dt$. Concisely, we write

$$\Gamma(dt) = N(0, 1/dt). \tag{13.73}$$

Note that Γ is then related to a random process called a Wiener process, which generates a normal random variable with zero mean and variance of dt. The Wiener process can be expressed as

$$dW(dt) = N(0, dt). \tag{13.74}$$

We can then write

$$dW(dt) = \Gamma(dt)dt. \tag{13.75}$$

The general Langevin equation is now

$$dX = f(X, t)dt + g(X, t)dW(dt). \tag{13.76}$$

Equation 13.76 is a stochastic differential equation that can describe the time evolution of variable X, which is under the influence of noise. Generally, the Langevin equation can be used to sample the probability distribution $P(X, t)$. Indeed it can be proven that the solutions of the Fokker–Planck equation and of the Langevin equation are equivalent (see Moyal's text in Further reading).

13.3.5 Chemical Langevin equations

Daniel Gillespie devised a Langevin equation that describes the time evolution of concentrations of species reacting away from the thermodynamic limit.

Consider a system of N chemical species reacting through M irreversible reactions. The chemical species state vector is $\underline{X} = (X_1, X_2, \ldots X_N)$, where X_i is the concentration of species S_i. The initial condition is taken to be $\underline{X}_o$. The $M \times N$ stoichiometric matrix $\underline{\underline{v}}$ has as elements the stoichiometric coefficients of species i in reaction j, v_{ij}.

For each of the chemical species S_i, Kuntz and Gillespie derived the following chemical Langevin equation (CLE):

$$dX_i = \sum_{j=1}^{M} v_{ij}a_j dt + \sum_{j=1}^{M} v_{ij}\sqrt{a_j}dW_j. \qquad (13.77)$$

The CLE is a multivariate Ito stochastic differential equation with multiple, multiplicative noise. We define the CLE again and present methods to solve it in Chapter 18, where we discuss numerical simulations of stochastic reaction kinetics.

13.4 Further reading

1. D. T. Gillespie, *Markov Processes*, (San Diego: Academic Press, 1992).
2. N. G. van Kampen, *Stochastic Processes in Physics and Chemistry*, (Amsterdam: North-Holland, 1981).
3. C. W. Gardiner, *Handbook of Stochastic Methods for Physics, Chemistry and the Natural Sciences*, (Berlin: Springer-Verlag, 1985).
4. H. Risken, *The Fokker–Planck Equation*, (Berlin: Springer, 1984).
5. I. Oppenheim and K. E. Shuler, *Phys. Rev. B*, **138**, 1007, (1965).
6. J. E. Moyal and J. Roy. *Statist. Soc. B*, **11**, 150, (1949).
7. D. A. McQuarrie, *J. Appl. Prob.*, **4**, 413, (1967).
8. D. T. Gillespie, *J. Chem. Phys.*, **113**, 297, (2000).

13.5 Exercises

1. Prove that $c_1 = k_1/V$ for reaction 13.24 and that $c_1 = 2k_1/V$ for reaction 13.25.

2. Starting with Eq. 13.33 prove Eqs. 13.34–13.37.

3. Prove Eq. 13.75.

4. Derive the chemical master equation for the reversible reaction in Eq. 13.1. Solve for the probability density and compare to the deterministic solution.

14

Molecular simulations

14.1 Tractable exploration of phase space

Statistical mechanics is a powerful and elegant theory, at whose core lie the concepts of phase space, probability distributions, and partition functions. With these concepts, statistical mechanics explains physical properties, such as entropy, and physical phenomena, such as irreversibility, all in terms of microscopic properties.

The stimulating nature of statistical mechanical theories notwithstanding, calculation of partition functions requires increasingly severe approximations and idealizations as the density of systems increases and particles begin interacting. Even with unrealistic approximations, the partition function is not calculable for systems at high densities. The difficulty lies with the functional dependence of the partition function on the system volume. This dependence cannot be analytically determined, because of the impossible calculation of the configurational integral when the potential energy is not zero. For five decades then, after Gibbs posited the foundation of statistical mechanics, scientists were limited to solving only problems that accepted analytical solutions.

The invention of digital computers ushered in a new era of computational methods that numerically determine the partition function, rendering feasible the connection between microscopic Hamiltonians and thermodynamic behavior for complex systems.

Two large classes of computer simulation method were created in the 1940s and 1950s:
1. Monte Carlo simulations, and
2. Molecular dynamics simulations.

Both classes generate ensembles of points in the phase space of a well-defined system. Computer simulations start with a microscopic model of the Hamiltonian. At equilibrium, the determination of the Hamiltonian is sufficient to define the geometry and size of the phase space. With the phase space constraints defined and remembering that macroscopic, thermodynamic properties of matter are nothing but ensemble averages over points in phase space, the objective of Monte Carlo and molecular dynamics simulations is the numerical generation of a large enough sample of the ensemble for the averages of thermodynamic properties to be estimated accurately.

On the one hand, molecular dynamics is based on integration of Newton's equations of motion, generating a succession of microscopic states on a phase space trajectory. On the other, Monte Carlo algorithms sample the phase space generating points with the help of random numbers. Remarkably, a sufficiently large number of points from either method results in the same thermodynamic property values. Although far from being a proof, this is in accord with and validates the ergodic hypothesis.

Is the calculation of the entire partition function not necessary then? The answer is that it is not necessary to calculate the actual value of the partition function, unless absolute values of the entropy and the free energies of a system are sought. We have discussed previously, in Chapter 5, that this is a chimeric endeavor and one that is ultimately unnecessary. There are at least two reasons for this. First, we are generally interested in changes of entropy and free energy during processes that change the thermodynamic state. Consequently there is no need for absolute values of these properties. The careful determination of a reference state is sufficient to compute relative entropies and free energies. Second, the absolute value of volumetric properties is derivable from the functional dependence of the partition function on system size, volume, or temperature. For example, the equation of state of an ideal gas is readily determined with only the realization that the partition function scales with V^N, without ever having to calculate what this value is.

Without the need to compute the total partition function, we can turn to numerical generation of statistically significant samples of the phase space so as to determine the thermodynamic behavior of matter, beginning with microscopic models.

In the next few chapters, we present computer simulations as a natural extension of statistical mechanical theories. We detail the methodological steps and provide a clear sequence of numerical steps, algorithms and computer codes to implement Monte Carlo and molecular dynamics simulations of simple systems. The reader will also benefit from numerous program codes made freely available on this book's sourceforge project at http://sourceforge.net/projects/statthermo. In this chapter we discuss prerequisite steps in all computer simulations.

We start, however, with a brief, parenthetic discussion of philosophical arguments in favor of computer simulations.

14.2 Computer simulations are tractable mathematics

The invention of differential and integral calculus in the 1660s was a remarkable accomplishment. Indeed, much of progress in the physical sciences can be credited to the pioneering work of Newton and Leibnitz. This is because calculus enhanced the unaided human brain in the following two major ways.

The first enhancement category is extrapolation, which is related to innate human computing capacities. This type of enhancement of human capacities is typically exemplified with instruments like the telescope or the microscope. These inventions enable visual detection beyond the range of the human eye. Similarly, the argument of a quantitative improvement in abilities can be made for calculus. As a methodological tool it extrapolates the capacities of the human brain. It is indeed difficult to imagine how the following integral could be defined, let alone determined, without calculus as a tool:

$$\int_{-\infty}^{+\infty} \exp(-p^2)dp = \sqrt{\pi}. \tag{14.1}$$

The second category of human capacity enhancement is augmentation. Augmentation is well exemplified with nuclear magnetic resonance instruments. There is no a-priori human ability to detect the resonance of nuclear magnetic moments to an external magnetic field. NMR equipment gives humans instrumental access to physical phenomena beyond our unaided capacities.

Analogously, calculus provides access to tractable mathematics and analytical solutions previously inaccessible to the human brain. Augmentation can then be considered as a qualitative shift in abilities. With results attainable only with calculus, the foundation can be solidly laid for theories that capture and explain physical phenomena. The development of the gravitational theory, the electromagnetic theory, or the quantum mechanical theory, is now possible, resulting, in turn, in tectonic changes in the human mindset.

Of course, with calculus analytical solutions became tractable only for linear, deterministic problems. For non-linear or probabilistic phenomena, the invention of computational mathematics has introduced an equivalently distinctive set of scientific methods. Paul Humphreys has best presented convincing arguments that computational science extrapolates and augments human understanding abilities in his wonderful book *Extending Ourselves* (see Further reading).

Physical systems are often non-linear or stochastic; they also often possess an overwhelming number of variables. Consequently, although in principle these systems can be described with the mathematical tools of calculus, in practice their behavior cannot be predicted or satisfactorily explained because of the intractability of analytical solutions. The determination of statistical mechanical properties is a strong case in point. There are insurmountable mathematical difficulties to develop analytical, predictive models of the thermodynamic properties of high density or multicomponent systems.

Computer simulation methods provide the much needed tractable mathematics. Because solutions are too complex, only computer models and simulations that are solidly founded on physical principles can extrapolate and augment the unaided human brain's capacities to describe, explain, and predict physical phenomena.

14.3 Introduction to molecular simulation techniques

Every computer simulation, whether molecular dynamics or Monte Carlo, starts with a clear definition of the following:

1. *A molecular model.* The Hamiltonian must be constructed with the relevant degrees of freedom and their interactions.
2. *Constraints.* The ensemble constraints, whether $NVE, NVT, \mu PT$, etc., must be defined.
3. *Geometry.* The size of the system and its geometric arrangement in space must also be defined.

We discuss each of these elements separately.

14.3.1 Construction of the molecular model

For non-charged, monoatomic systems we assume pairwise additivity of interactions. We then write for the potential energy function

$$U(\underline{r}) = \sum_{i>j} u(r_{ij}). \tag{14.2}$$

The pairwise interaction potential can be described by a Lennard-Jones potential,

$$u(r_{ij}) = 4\varepsilon \left(\left(\frac{\sigma}{r_{ij}} \right)^{12} - \left(\frac{\sigma}{r_{ij}} \right)^{6} \right). \tag{14.3}$$

Table 14.1 *Lennard-Jones interaction parameters.*

Atom	$\sigma(\text{Å})$	ε/k_B (K)
Hydrogen	2.81	8.6
Helium	2.28	10.2
Carbon	3.35	51.2
Nitrogen	3.31	37.3
Oxygen	2.95	61.6
Fluorine	2.83	52.8
Argon	3.41	119.8

In Eq. 14.3, σ is indicative of the size of the atoms, so that $u(\sigma) = 0$, whereas ε is the maximum depth interaction energy, so that $u(r_{ij}) = \varepsilon$ when $\partial u(r_{ij})/\partial r_{ij} = 0$.

In Table 14.1, interaction parameters are shown for some atoms (see the book by Allen and Tildesley for parameters of more atoms).

For unlike atoms, the Lorentz–Berthelot combining rules are used to determine the Lennard-Jones parameters. These rules are as follows:

$$\sigma_{12} = \frac{1}{2}(\sigma_1 + \sigma_2) \tag{14.4}$$

and

$$\varepsilon_{12} = \sqrt{\varepsilon_1 \varepsilon_2}, \tag{14.5}$$

where σ_1, ε_1 and σ_2, ε_2 are the Lennard-Jones parameters of species 1 and 2, respectively.

For molecular species we need more terms to model the internal degrees of freedom and their interactions. Typically the additional energy terms are associated with covalent bonds, bond angles, and dihedral angles. We consider systems with M molecules with N_M atoms each for a total of $N = M \times N_M$ atoms in the system.

Bond length potential

The energy of a covalent bond between two atoms that are separated by a distance r can be modeled with a harmonic potential,

$$V_r = \frac{1}{2}k_r(r - r_o)^2, \tag{14.6}$$

where r_o is the equilibrium bond length and k_r is the spring constant.

A more accurate model is the Morse potential, although it is not used as frequently in classical computer simulations. The functional form of

Table 14.2 *Example bond equilibrium lengths and energies.*

Atom pair	r_o (Å)	k_r (kcalmol^{-1}Å^{-2})
C-C	1.523	317
C=C	1.357	690
C-N	1.449	337
N-H	1.010	434

the Morse potential is the following:

$$V_r = E_r \left(1 - e^{-a(r-r_o)}\right)^2, \qquad (14.7)$$

where E_r is the potential energy for bond formation and a is a potential well width parameter.

The harmonic approximation is valid as long as the thermal energy is lower than the energy of bond formation, $k_B T < E_r$.

It is worth noting that when the bond length is at its equilibrium value, $r = r_o$, the potential energy in Eq. 14.6 is zero. The energy of forming or breaking the bond is certainly not zero, but in classical mechanical computer simulations covalent bonds remain intact. In order to model and simulate phenomena involving bond formation or breakage, a quantum mechanical treatment would be necessary. Descriptions of such a treatment are beyond the scope of this text. The interested reader is referred to references in Further reading.

In Table 14.2 values are shown of representative bond energies and equilibrium lengths.

Bond angle potential

The bond angle, $A - B - C$, between three atoms is defined as the angle between the bonds $A - B$ and $B - C$ (Fig. 14.1). A harmonic potential is typically used to model bond angle energies,

$$V_\theta = \frac{1}{2} k_\theta \left(\theta - \theta_o\right)^2, \qquad (14.8)$$

where θ_o is the equilibrium bond angle.

In Table 14.3, typical values of bond angle energies are shown.

Dihedral or torsional angles

The dihedral or torsional angle ϕ between four atoms $A - B - C - D$ is defined as the angle between the two planes formed by atoms ABC and

Table 14.3 *Example equilibrium bond angles and bond angle energies.*

Angle	θ_o (deg)	k_θ (kcalmol^{-1}deg^{-2})
C-C-C	109.47	70
H-C-H	109.47	40
C-N-H	120.00	30
H-O-H	104.52	55

by atoms BCD. Typically, the dihedral angle potential V_ϕ is expressed as an expansion in ϕ. For a linear molecule with N_M atoms we can write

$$V_\phi = \sum_{i=2}^{N_M-2} \sum_{j=0}^{5} c_j \left(\cos\phi_j\right)^j, \tag{14.9}$$

where c_j are fitting coefficients. The number of terms in the second sum is chosen to truncate accurately a Fourier series with infinite terms. Six terms is a typical number of terms used in classical mechanical simulations.

Electrostatic interactions

Electrostatic interactions, such as ionic interactions, are modeled with Coulomb's law

$$V_{\text{ele}} = \sum_{i=1}^{N_c} \sum_{j=i+1}^{N_c-1} \frac{q_i q_j}{4\pi\varepsilon_o r_{ij}}, \tag{14.10}$$

where N_c is the number of charged atoms in the system.

Figure 14.1 Simple molecular motion types. Bond length, bond angle, and dihedral angle motions.

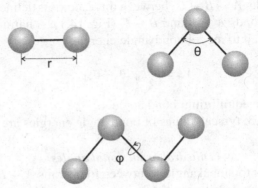

Atoms involved in covalent bonds can also carry partial charges because of differences in their electronegativity. For example, the oxygen atom in a water molecule attracts the electrons away from the hydrogen atoms, because of its high electronegativity. This results in an apparent partial negative charge on the oxygen and a partial positive charge on each of the hydrogens. Interactions between partial charges are again modeled with Coulomb's law.

There can also be higher order electrostatic interactions such as charge–dipole and dipole–dipole interactions. A discussion of these types of interaction is beyond the scope of this text.

14.3.2 Semi-empirical force field potential

The sum of all non-bonded (Lennard-Jones and electrostatic) and bonded (bond length, bond angle, dihedral angle) interactions is the potential energy of the system, also called a force field potential. Compactly, we write for the total potential energy of the system the following expression:

$$U(\underline{r}) = U_{\text{non-bonded}} + U_{\text{bonded}}, \qquad (14.11)$$

with $U_{\text{non-bonded}} = U_{\text{Lennard-Jones}} + U_{\text{electrostatic}}$ and $U_{\text{bonded}} = U_{\text{bonds}} + U_{\text{bondangles}} + U_{\text{torsionangles}}$, where U_{bonds} is the sum of all the bond energies in the system, $U_{\text{electrostatic}}$ is the sum of all electrostatic interactions, etc.

A typical classical mechanical potential energy for a system of M linear molecules, with N_M atoms each, for a total of N atoms in the system ($N = M \times N_M$), is then written as

$$U(\underline{r}) = \sum_{j=1}^{M} \sum_{i=1}^{N_M-1} \frac{k_{rij}}{2}(r_{ij} - r_{ijo})^2 + \sum_{j=1}^{M} \sum_{i=1}^{N_M-2} \frac{k_{\theta ij}}{2}(\theta_{ij} - \theta_{ijo})^2$$
$$+ \sum_{j=1}^{M} \sum_{i=1}^{N_M-2} \sum_{k=0}^{5} c_k(\cos\phi_{ijk})^k$$
$$+ \sum_{i=1}^{N} \sum_{j=i+1}^{N-1} \left(4\varepsilon_{ij} \left[\left(\frac{\sigma_{ij}}{r_{ij}}\right)^{12} - \left(\frac{\sigma_{ij}}{r_{ij}}\right)^{6} \right] + \frac{q_i q_j}{4\pi\varepsilon_o r_{ij}} \right).$$
$$(14.12)$$

This potential is semi-empirical because some terms and parameters are determined using quantum mechanical principles, while most are determined empirically so that a good fit is achieved with available experimental observations. The process of determining the force field parameters is a tedious one that requires extensive validation between simulated and experimentally measured properties. Peter Kollman, William Jorgensen, and Martin Karplus are but a few of the scientists who have pioneered the development of force fields for simulations.

There are numerous available classical force field potentials, such as CHARMM, AMBER, GROMOSS, OPLS, MMFF, CVFF, and TRAPE. They each have particular advantages and disadvantages, largely dependent on the system studied. Some, like AMBER and CHARMM, provide accurate parameters for biological systems, others such as OPLS and MMFF offer accurate parameters for small organic molecules, etc. Some, such as CHARMM and AMBER, are also available as parts of large software molecular simulation packages that include routines for molecular dynamics or Monte Carlo simulations.

It is important to note that however powerful they may be, empirical force field potentials are not perfectly accurate under all possible conditions. For example, a water molecule in the widely used TIP3P model has such partial charges on the oxygen and hydrogen atoms that result in an equilibrium constant dipole moment of 2.25 Debye. This is close to the effective dipole moment of around 2.5 Debye of liquid water at room temperature. On the other hand a single water molecule in vacuum has a dipole moment of 1.85 Debye. The difference between the bulk and the gas water dipole moment is the result of high polarizability at higher densities. TIP3P and most other classical mechanical force fields do not account for polarizability.

Polarizable force fields

In standard force field potentials the distribution of partial charges on atoms is fixed, with the charges located on atom center-of-mass. These distributions are often determined from quantum mechanical calculations of molecules in vacuum, or in a mean-field representation of the environment of the molecule. In real physical systems, especially at high densities, molecules polarize substantially. There have been numerous polarizable force fields developed, where the charge at each atomic center can change depending on its environment. For example, Matsuoka and co-workers (see Further reading) developed a polarizable water force field potential. The advantage is that such potential is transferable to varying environments. The disadvantage is that there is a need to fit ten potential parameters. Calculation of the system potential energy is also more tedious and time consuming.

Potential truncation

Calculation of the non-bonded empirical potential energy requires sums in Eq. 14.2 that may in principle extend to infinite distances. For the calculation to be tractable, we can truncate the potential, by picking a distance cutoff, r_c, such that

$$u(r_{ij}) = 0, \quad \text{for } r_{ij} > r_c. \tag{14.13}$$

The cutoff distance is typically chosen to be $r_c \approx 2.5\sigma$. At this distance the dispersion force between any two atoms is indeed close to zero.

Simply adopting Eq. 14.13 results in a discontinuity of the potential at r_c, i.e.,

$$\frac{du}{dr} = \infty. \quad r = r_c \tag{14.14}$$

This discontinuity may result in numerical instabilities during a simulation, because the value of the force may be computed to be much larger than the largest number understood by the computer. In order to avoid the discontinuity the entire potential can be shifted by a constant $u(r_c)$. As a result, the Lennard-Jones interaction energy becomes zero at r_c in a smooth, continuous way (Fig. 14.2). A long-range correction is then needed, because the potential is actually not zero at the cutoff

Figure 14.2 Truncated Lennard-Jones potential.

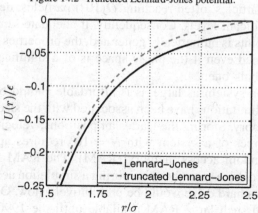

distance. Assuming that

$$g(r) = 1, \quad \text{if } r_{ij} > r_c \tag{14.15}$$

we can add the following truncation correction term to the total Lennard-Jones energy of a system:

$$2\pi N \int_{r_c}^{\infty} r^2 u(r) g(r) dr. \tag{14.16}$$

Whereas dispersion forces are straightforward to model, this is not the case with electrostatic interactions. Coulombic interactions act at significantly longer distances, falling off with r^{-1} instead of r^{-6}, and require special treatment, with a method called Ewald summation. A discussion of Ewald's method is beyond the scope of this book, and the interested reader is referred to the text by Allen and Tildesley (Further reading).

14.3.3 System size and geometry

The argument that even the most powerful digital computer cannot compute the potential energy of a system with macroscopic dimensions was already discussed in the first chapter. It is not feasible to determine properties of Avogadro's number of particles. Thankfully, it is also unnecessary.

The fluctuations of thermodynamic properties computed from microscopic systems of size N scale as $(\sqrt{N})^{-1}$. With $N = 10\,000$ particles, the standard deviation of a macroscopic state function is already 1% of the property value. Currently, simulations can be routinely conducted with $O(10^5)$ particles, often reaching $O(10^6)$ particles, decreasing the error to insignificant levels. Consequently, if adequate sampling of the phase space points is numerically generated, the properties can be accurately determined even if the phase space is of a significantly smaller system than a bulk one.

Practically, a system as large as is affordable should be simulated. Traditionally, limitations have been associated with the size of available computer memory. Storing the Cartesian positions, velocities, accelerations, and forces of a system with $N = 10^5$ requires approximately 100 MB of random access memory (RAM). The RAM is needed as opposed to hard drive space because during a simulation accessing these numbers from a hard drive would be prohibitively slow. Only high-end computers had such large RAM available until the 1990s. Currently there is ample, inexpensive computer memory available to conduct sufficiently large simulations of simple liquids even on desktop computers. Consequently, computer memory is no longer as major a consideration when developing algorithms for computer simulations as it used to be.

Of consideration remains the actual wall-clock time required in order to generate a sufficiently large sample of microscopic states. We discuss the time-related considerations separately for Monte Carlo and molecular dynamics simulations in the next two chapters.

Perhaps the most important disadvantage of a finite size system with walls bounding the available volume is that a lot of particles will be located at or near the wall surface. Consequently, the properties obtained from the simulation will not be the bulk properties of the system. An ingenious solution to the challenges of simulating small system sizes is based on periodic boundary conditions, presented next.

14.3.4 *Periodic boundary conditions*

With periodic boundary conditions we simulate small systems in the bulk state (Fig. 14.3). The trick is to imagine surrounding the primary volume cell with images of itself in all directions. If a particle moves and leaves the central box, it gets replaced by one of its images in the

Figure 14.3 Periodic boundary conditions.

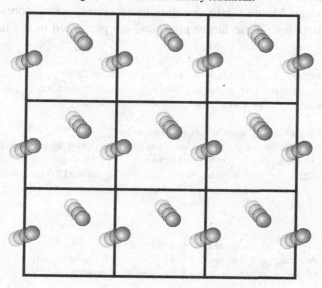

opposite side of the box. Consequently there are always N particles in the simulated system. Particles on or near the borders of the available volume do not interact with a surface. Instead they interact with the images of particles on the opposite side of the simulated system. In order to appreciate the concept of periodic boundary conditions, it is best to describe the actual algorithm that implements periodic boundary conditions in a computer program.

14.3.5 *FORTRAN code for periodic boundary conditions*

The periodic boundary conditions are implemented with the following FORTRAN piece of code. FORTRAN has traditionally been the computer language of choice for simulation algorithms. A simple reason is that FORTRAN compilers, the programs that translate the user-written code to machine language understood by the computer, have been historically optimized for scientific computation. The syntax and commands of FORTRAN are arguably fairly intuitive. In particular there are only a few commands that FORTRAN codes are largely written with. With some reasonable effort the reader could learn those and should manage to at least read through the FORTRAN algorithms presented in this textbook. A primer of simple FORTRAN commands is provided in the textbook's webpage at http://sourceforge.net/projects/statthermo.

Here is the FORTRAN code that implements the periodic boundary conditions. This is only part of the larger programs for molecular dynamics and Monte Carlo simulations of simple monoatomic fluids with N particles. These larger programs are presented in the next two chapters.

```
c Any FORTRAN program line starting with a c is
c not executed by the program. Such lines are used to
c add comments that make the code more readable.

c Start by defining system size variables.
c V is the volume of the system, assumed to have been determined
c prior to this point in the program.
c boxl is the length of the box, and boxl2 is half the box size.

boxl=V**(1/3)
boxl2  =  boxl/2

c Start a do-loop over all N particles.
c This next command is used to iterate and repeat executions.
c do-loops in FORTRAN end with an enddo command.
```

```
do i=1 to N

c The next three lines implement the periodic boundary conditions.
c If a particle moves outside the box, its image enters the box
c in the opposite side.

if  (rx(i).gt.boxl2) then rx(i) = rx(i) - boxl
if  (rx(i).lt.-boxl2) then rx(i) = rx(i) + boxl
if  (ry(i).gt.boxl2) then ry(i) = ry(i) - boxl
if  (ry(i).lt.-boxl2) then ry(i) = ry(i) + boxl
if  (rz(i).gt.boxl2) then rz(i) = rz(i) - boxl
if  (rz(i).lt.-boxl2) then rz(i) = rz(i) + boxl

c End the loop

enddo
```

The implementation of the periodic boundary conditions will be clear with an example, presented in the next section.

We should stress that programming algorithms for computer simulations is an art. There is no one single way to implement an algorithm with a programming language. A skilled programmer is always looking out for computational gains. In the previous example, executing the "if" statement $6N$ times for every generated point in phase space becomes a computationally intensive task. Instead one can better use the "anint" function of FORTRAN. Calling "anint(x)" for any variable x returns the nearest integer to x. For example, "anint(0.49) = 0", and "anint(0.51) = 1."

We replace the "if" statements in the loop to write

```
rx(i) = rx(i) - boxl*anint(rx(i)/boxl)
ry(i) = ry(i) - boxl*anint(ry(i)/boxl)
rz(i) = rz(i) - boxl*anint(rz(i)/boxl)
```

Although this piece of code will result in exactly the same results as the original one, it will result in a considerably faster simulation, requiring a shorter wall-clock time to complete the computations.

14.3.6 Minimum image convection

When periodic boundary conditions are used, the system must be large enough for particles with their own images. For non-charged systems, this is accomplished by having the simulated box be larger than twice the Lennard-Jones potential cutoff. Each atom experiences then at most one image of every other atom in the system. Charged systems require special treatment because of the long range of the electrostatic potential, discussed in references in Further reading.

The minimum image convention (Fig. 14.4) is used to ensure that a particle interacts with another particle or only one of its images, whichever is closest. A piece of FORTRAN code for computing the potential energy of a system with N particles interacting with a simple Lennard-Jones potential is as follows:

```
c This code loops over all pairs of particles to determine
c their interactions and the total energy.
c The positions rx(i), ry(i), rz(i) of all particles are known.
c In the beginning, it is a good idea to zero out a variable
c like the potential energy U.

U =  0.0d0

c Start the first do-loop to compute all interactions

do  i=1, N-1

c It is faster for the computer to use scalar variables
c instead of vector variables.
c Since each position  variables rx(i), ry(i), rz(i)
c will be used multiple times in the
c next loop over j particles, it is better then to
```

Figure 14.4 Minimum image convention.

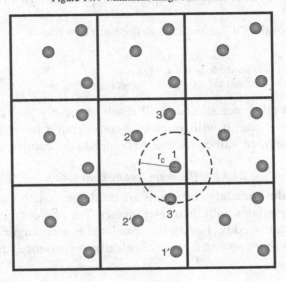

```
c define new scalar variables and pass
c the current values of the vector variable to them.

rxi = rx(i)
ryi = ry(i)
rzi = rz(i)

c Open the second loop

do j=i+1, N

c Compute the distance between particles i and j

rxij = rxi - rx(j)
ryij = ryi - ry(j)
rzij = rzi - rz(j)


c Apply the minimum image convention so
c that each particle i interacts with the closest image of j

rxij = rxij - boxl*anint(rxij/boxl)
ryij = ryij - boxl*anint(ryij/boxl)
rzij = rzij - boxl*anint(rzij/boxl)

c Next we compute the distance rij

rij = sqrt(rxij**2 + ryij**2 + rzij**2)

c Determine the Lennard-Jones interaction between i and j

srij = sigma/rij
srij = srij**6

ULJ = 4 * epsilon *(srij + srij**2)

c End do-loop over j particles

enddo

c Add the Lennard-Jones interaction potential to the total
c potential energy.

U = U + ULJ

c End do-loop over i particles

enddo
```

Example 14.1

To illustrate the use of the periodic boundary conditions and the minimum image convention, let us consider a system with three particles, as shown in Fig. 14.5, in a box of length $L = 5$, measured in reduced units so that $\sigma = 1$ (this can be achieved by dividing all lengths by the size of the particles. For simplicity, let us assume that all three particles are at the same z-position, $r_z = 0$. The position coordinates are then $(rx(1), ry(1), rz(1)) = (1.5, -1.5, 0)$, $(rx(2), ry(2), rz(2)) = (0, -1.5, 0)$, and $(rx(3), ry(3), rz(3)) = (1.5, 1.5, 0)$. Assume that the cutoff of the Lennard-Jones potential is $U_c = 2.4$.

In the previous code, the first value for i is $i = 1$ and according to the figure positions:

```
rxi = rx(1) = 1.5
ryi = ry(1) = -1.5
rzi = rz(1) = 0
```

The first value of j is $j = i + 1 = 2$. Then

```
rxij = rxi - rx(2) = 1.5 - 0 = 1.5
ryij = ryi - ry(2) = -1.5 - (-1.5) = 0
rzij = rzi - rz(2) = 0 - 0 = 0
```

The closest image of particle 2 to particle 1 is the one inside the main cell. The minimum image convention will then give the same distances as before.

Figure 14.5 Three particles interacting through periodic boundaries with the minimum image convention.

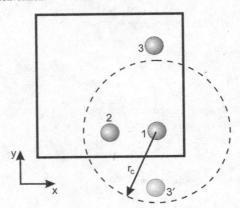

```
rxij = rxij - boxl*anint(rxij/boxl) = 1.5 - 5*anint(1.5/5) = 1.5 -
   0 = 1.5
ryij = ryij - boxl*anint(ryij/boxl) = 0 - 5*anint(0/5) = 0 - 0 = 0
rzij = rzij - boxl*anint(rzij/boxl) = 0 - 5*anint(0/5) = 0 - 0 = 0
```

The code proceeds to determine the distance between particles 1 and 2, compute the Lennard-Jones interaction and add it to the potential energy.

The do-loop over the j variable then returns. The variable i is still $i = 1$ and j becomes $j = 3$.

The position variables for i retain the same values at this point:

```
rxi = rx(1) = 1.5
ryi = ry(1) = -1.5
rzi = rz(1) = 0
```

With $j = 3$ the code then returns:

```
rxij = rxi - rx(3) = 1.5 - 1.5 = 0
ryij = ryi - ry(3) = -1.5 - 1.5 = -3
rzij = rzi - rz(3) = 0 - 0 = 0
```

This time, the closest image of particle 3 to particle 1 is not the one inside the main cell. Instead it is the one in the lower image cell. The minimum image convention will then give a different distance between particles 1 and 3:

```
rxij = rxij - boxl*anint(rxij/boxl) = 0 - 5*anint(0/5) = 0 - 0 = 0
ryij = ryij - boxl*anint(ryij/boxl) = -3 - 5*anint(-3/5) = -3 - 5*(-1) = 2
rzij = rzij - boxl*anint(rzij/boxl) = 0 - 5*anint(0/5) = 1 - 0 = 0
```

The distance between particles 1 and 3 used in the Lennard-Jones calculation is the closest image one, in this case calculated to be $r_{13} = 2$, in reduced units. This is less than the Lennard-Jones cutoff. Particle 1 then does interact with the image of particle 3, through the periodic boundaries, and another term is added to the total energy of the system.

After calculation of the potential energy the algorithm will generate a new configuration of particle positions. For example, if the simulation is a molecular dynamics one, the program will sum all the forces on each particle calculated with the Lennard-Jones potential, divide the total force by the particle mass to determine the acceleration, and propagate numerically the positions of the particles. We discuss how this is done in the next two chapters.

For the sake of the discussion, let us assume that the new positions of particles 1, 2, and 3, generated by a single Monte Carlo or molecular dynamics simulation step, are as shown in Fig. 14.6. Particles 2 and 3 are still inside the box in this new configuration, but particle 1 has moved outside the box to

$$(rx(1), ry(1), rz(1)) = (3, -1, 0).$$

Before the loops return to the beginning of the program, periodic boundary conditions are applied:

```
do i=1, 3
rx(i) = rx(i) - boxl*anint(rx(i)/boxl)
ry(i) = ry(i) - boxl*anint(ry(i)/boxl)
rz(i) = rz(i) - boxl*anint(rz(i)/boxl)
enddo
```

For particles 2 and 3 their positions remain unchanged by this piece of code. For particle 1 the new position is found to be $(rx, ry, rz) = (-2, -1, 0)$. This virtually brings an image of particle 1 in the box. This image has now become particle 1 and the number of particles inside the box remains constant.

14.4 How to start a simulation

In Monte Carlo simulations, an initial condition must be generated at the start of the simulation, defining positions for all particles. In molecular dynamics simulations, besides the positions, the initial velocities must be defined.

The necessary amount of care for generating the initial condition $(\underline{r}(0), \underline{v}(0))$ of a simulated system depends on the relevant time scales of interesting motions. For a simple fluid of Lennard-Jones particles at room temperature, the particles move fast enough for the system to quickly reach equilibrium regardless of the initial configuration. Consequently, the simplest-to-build configuration may be chosen. This could be, for example, a simple FCC lattice with the particles originally placed

Figure 14.6 Application of periodic boundary conditions brings all particles inside the simulation box.

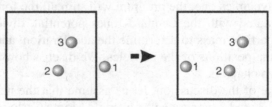

on the nodes of the lattice. A short simulation will then result in the equilibrium fluid structure. This can be monitored with a computed pair distribution function, as discussed later in Chapter 15.

In molecular dynamics, the initial condition must include the velocities. These can be drawn from a Maxwell–Boltzmann distribution,

$$f(\underline{u}) = \left(\frac{m}{2\pi k_B T}\right)^{3/2} \exp\left(-\frac{mu^2}{2k_B T}\right). \tag{14.17}$$

Care must be taken so that the total momentum of the system in all directions is zero. This can be achieved by choosing the velocity of the last particle, so that

$$\sum_{i=1}^{N} m_i \underline{u}_i = 0. \tag{14.18}$$

On the other hand, for complex systems it is imperative to generate an initial condition that is as close to the equilibrium structure as possible. For example, if a mixture of water and alcohol molecules are simulated at room temperature, the molecules may be best randomly placed at the beginning of the simulation. Since the two liquids mix well, a random initial configuration may be preferable to a lattice one with the molecular phases separated.

In principle, if any simulation is conducted for long enough, the equilibrium state of the system will be reached eventually. Nonetheless, there may be instances where the system is initiated so far from equilibrium that inordinate computational resources may be required to reach equilibrium. Indeed, there may be systems for which no practically attainable simulation time can result in equilibrium states. As an example consider a protein molecule. A typical protein attains a particular three-dimensional conformation at equilibrium. The protein folds from an extended conformation to its equilibrium structure in a time interval that ranges from milliseconds to minutes, largely depending on its size. Though longer time scales become constantly available with the help of ever more advanced supercomputing resources, currently there is no capacity to simulate a fully solvated protein for more than a few microseconds. This is not enough to reach equilibrium, if the initial conformation is chosen very far from the equilibrium conformation. The choice of initial configuration may then make the difference between feasible and intractable simulations.

14.5 Non-dimensional simulation parameters

The calculations in a molecular simulation program may be considerably simplified using reduced, or non-dimensional, simulation units. For example, in a simulation of argon, using $m = 39.94800$ amu, let alone $m = 6.63352065 \times 10^{-26}$ kg, would complicate the numerical calculations unnecessarily. A way to simplify the calculations is by a suitable choice of parameters that will reduce the units. Typically the following four parameters are chosen to non-dimensionalize simulation calculations: the particle mass, m, the particle size, σ, the particle charge, q, and the potential well minimum, ε. All mass terms are then scaled by m, all length terms are scaled by σ, all charge terms are scaled by q, and all energy terms are scaled by ε.

In a simulation with only one atom type, all these terms are then one. Consequently, all calculations are substantially simplified (for example, in the calculation of the Lennard-Jones interaction, $\varepsilon^* = 1$ and $\sigma^* = 1$). Indeed a major benefit is that with non-dimensional parameters all numerical values in a simulation are of order magnitude $O(1)$.

Other parameters are also scaled appropriately. For example, the following reduced parameters can be used:

$$\rho^* = \rho \sigma^3, \tag{14.19}$$
$$T^* = k_B T / \varepsilon, \tag{14.20}$$
$$U^* = U / \varepsilon, \tag{14.21}$$
$$P^* = P \sigma^3 / \varepsilon, \tag{14.22}$$
$$t^* = t \left(\varepsilon / m \sigma^2 \right), \tag{14.23}$$

and so on.

Another benefit of using reduced units is that a single simulation of a Lennard-Jones fluid pertains to different actual system simulations. This then becomes an interesting illustration of the law of corresponding states.

14.6 Neighbor lists: a time-saving trick

In the inner loops of both Monte Carlo and molecular dynamics algorithms we consider particle i and loop over all other particles j to calculate the distance and the minimum image separations. If the distance is less than the potential cutoff distance, the interaction potential and the forces are computed. In this manner, all pairs of particles are examined

Table 14.4 *Indicative values of list size and update frequency.*

List radius	Update interval	CPU time(s) $N = 256$	CPU time(s) $N = 512$
no list	–	3.33	10
2.6 σ	5.78	2.24	4.93
2.7 σ	12.5	2.17	4.55
2.9 σ	26.32	2.28	4.51
3.10 σ	43.48	2.47	4.79

at every iteration. Consequently, the required simulation wall-clock time scales with the square number of particles, N^2.

Loup Verlet proposed a simple yet brilliant technique for improving the speed of computer simulations. In Verlet's algorithm, a list is maintained of the neighbors of any particular particle i. The program calculates only the distance and the interaction potential between particle i and its neighbors, neglecting the rest of the particles. The neighbors are defined as the particles a certain distance away from particle i. The distance is somewhat larger than the potential cutoff distance of 2.5σ, usually between 2.6 and 3.0σ.

The neighbor list is updated at regular intervals. The update frequency depends on the neighbor list size. The shorter the distance classifying neighbors, the smaller the neighbor list. The update frequency is then high. If the neighbor list is large, the update frequency need not be high. The list size and update frequency are determined empirically. Practically, this means that a few test simulations need to be conducted, in order to determine the optimum combination of neighbor list size and update frequency.

For Lennard-Jones fluids, Table 14.4 gives indicative values for list sizes and update frequencies for systems of two different sizes, with $N = 256$ and $N = 512$ particles. The CPU time is shown that is required for calculation of the total potential energy for the different cases. As observed, there is an optimum list radius and update frequency. Notably, the optimum depends on the size of the system.

14.7 Further reading

1. M. P. Allen and D. J. Tildesley, *Computer Simulation of Liquids*, (New York: Oxford University Press, 1989).
2. O. Matsuoka, E. Clementi, and M. J. Yoshimine, *J. Chem. Phys.*, **64**, 1351, (1976).

3. D. Frenkel and B. Smit, *Understanding Molecular Simulation*, 2nd ed., (San Diego: Academic, 2002).
4. D. N. Theodorou and U. W. Suter, *J. Chem. Phys.*, **82**, 955, (1985).
5. P. Humphreys, *Extending Ourselves*, (Oxford: Oxford University Press, 2004).

14.8 Exercises

1. Write a program that generates random positions for N particles in a box of volume V.

2. Write a program that generates random velocities for N particles at temperature T.

3. What is the total potential energy of the system in Fig. 14.5?

4. What will the result be of minimum image convention calculations, if the position of particle 2 in Fig. 14.5 is $(3, -1, 0)$?

5. Consider a simulation conducted with reduced units at $\rho = 0.4$ and $T = 1.2$. What conditions are modeled for argon, oxygen, and methane?

15

Monte Carlo simulations

Monte Carlo methods are computational techniques that use random sampling. Nicholas Metropolis and Stanislaw Ulam, both working for the Manhattan Project at the Los Alamos National Laboratory in the 1940s, first developed and used these methods. Ulam, who is known for designing the hydrogen bomb with Edward Teller, invented the method inspired, in his own words, "... by a question which occurred to me in 1946 as I was convalescing from an illness and playing solitaires. The question was what are the chances that a Canfield solitaire laid out with 52 cards will come out successfully?".* This was a new era of digital computers, and the answer Ulam gave involved generating many random numbers in a digital computer.

Metropolis and Ulam soon realized they could apply this method of successive random operations to physical problems, such as the one of neutron diffusion or the statistical calculation of volumetric properties of matter. Metropolis coined the term "Monte Carlo" in reference to the famous casino in Monte Carlo, Monaco, and the random processes in card games. Arguably, this is the most successful name ever given to a mathematical algorithm.

Nowadays Monte Carlo refers to very many different methods with a wide spectrum of applications. We present the Metropolis Monte Carlo method for sampling the phase space to compute ensemble averages, although often we only use the term "Monte Carlo" in subsequent discussions.

---

* Richard Canfield was a famous gambler in the late nineteenth century, considered America's first casino king. His story is well beyond the scope of this book.

15.1 Sampling of probability distribution functions

The canonical ensemble average of any thermodynamic property of matter, $M(\underline{p}, \underline{q})$, is

$$\langle M \rangle_{NVT} = \frac{\int_{\Gamma} M(\underline{p}, \underline{q}) \exp(-\beta H(\underline{p}, \underline{q})) d\underline{p} d\underline{q}}{\int_{\Gamma} \exp(-\beta H(\underline{p}, \underline{q})) d\underline{p} d\underline{q}}. \tag{15.1}$$

For example, the excess internal energy for non-ideal systems is the ensemble average of the potential energy

$$
\begin{aligned}
U^{\text{excess}} &= \langle U(\underline{X}) \rangle \\
&= \int_{\Gamma} U(\underline{X}) \rho(\underline{X}) d\underline{X} \\
&= \frac{\int_{\Gamma} U(\underline{p}, \underline{q}) \exp(-\beta H(\underline{p}, \underline{q})) d\underline{p} d\underline{q}}{\int_{\Gamma} \exp(-\beta H(\underline{p}, \underline{q})) d\underline{p} d\underline{q}},
\end{aligned}
\tag{15.2}
$$

where $\rho(\underline{X})$ is the probability density of any point $\underline{X} = (\underline{p}, \underline{q})$ in phase space Γ.

Remembering that the integral over the phase space is an approximation of the sum over phase space points, we can write

$$
\begin{aligned}
U^{\text{excess}} &= \sum_i U(\underline{X}_i) \rho(\underline{X}_i) \\
&= \frac{\sum_i U(\underline{X}_i) \exp(-\beta U(\underline{X}_i))}{\sum_i \exp(-\beta U(\underline{X}_i))},
\end{aligned}
\tag{15.3}
$$

where the sums run over all points in the configurational part of phase space.

For dense systems it is not possible to determine the integrals in Eq. 15.2, or the sums in Eq. 15.3 because the total number of the configurational phase space points representing an NVT system at the thermodynamic limit is stupendously large. Thankfully, as in any other statistical average, it is sufficient to sample the phase space and estimate the ensemble average of thermodynamic properties. If the phase space is sampled sufficiently well, the ratio of the sums in Eq. 15.3 can converge to the correct ensemble average value, even if the actual number of points sampled is much smaller than the entire ensemble.

One conceptually simple method to sample the phase space is to uniformly generate random points in it. Imagine throwing darts which land randomly on points $\underline{X}$ in the configurational part of phase space. In the following section we discuss how to implement uniform sampling of the phase space. We demonstrate that, although conceptually simple, this method is inefficient because it samples important and unimportant parts of the phase space alike. We argue that uniformly random sampling ultimately fails for high-density systems and we set the stage for the Metropolis Monte Carlo method. Metropolis Monte Carlo is an importance sampling method, generating points in areas of phase space in proportion to the probability density of these areas, $\rho(\underline{X})$. This is the secret of the success of the method Metropolis and Ulam developed.

15.2 Uniformly random sampling of phase space

Consider a model NVT system of particles interacting with the Lennard-Jones potential. We assume that periodic boundary conditions are used and that N is in the order of 10^5. We also assume that there is an available program subroutine that returns a uniform random number ζ in the interval $(0, 1)$ when called in the main simulation program.

The steps of an algorithm used to estimate the average excess energy by randomly generating N_{trials} points in phase space are as follows:

1. Generate $3N$ random numbers ζ in the interval $(0, 1)$.
2. Obtain a point $\underline{X}$ in configurational phase space by rescaling all ζ to compute random positions inside the system's volume.

   Specifically, for each particle determine its x position in a cubic simulation box of size L using the following expression:

   $$x = \zeta \times L - L/2.0, \tag{15.4}$$

   repeating for y and z with a different ζ each time. The origin of the coordinate system is assumed to be at the center of a cubic box.

   Drawing $3N$ different ζ, a random configuration is now generated for all N particles representing a point $\underline{X}$ in $3N$ configurational space.
3. Determine the potential energy, $U(\underline{X})$, of this configuration using periodic boundary conditions and the minimum image convention.
4. Calculate the Boltzmann factor, $\exp(-\beta U(\underline{X}))$.
5. Return to the first step and repeat for N_{trials} times.

6. Compute the ensemble average

$$U^{\text{excess}} = \langle U(\underline{X}) \rangle$$

$$= \frac{\displaystyle\sum_{\lambda=1}^{N_{\text{trial}}} U(\underline{X}_\lambda) \exp(-\beta U(\underline{X}_\lambda))}{\displaystyle\sum_{\lambda=1}^{N_{\text{trial}}} \exp(-\beta U(\underline{X}_\lambda))}. \tag{15.5}$$

The ensemble average of any other macroscopic property can be calculated accordingly.

With available high-performance computers a single calculation of the potential energy and the Boltzmann factor for a system with $N = 10^5$ can be conducted within a few minutes. Consequently, this is not a challenging step in the algorithm. Instead the difficulty is associated with the required number of generated points, N_{trials}. For dense systems, an inordinate number of trials may be required to converge to the correct ensemble average value, simply because the overwhelming majority of these generated points contribute nothing to the average in the last step.

For dense systems, chances are that configurations generated randomly will be of very high energy with a very small Boltzmann factor. To illustrate this point, imagine that the position of $N - 1$ particles has been generated and the algorithm proceeds to generate the position of the Nth particle. Even if somehow serendipitously the $N - 1$ previous particles have been positioned without strong interactions, the random position of the last particle will most probably be close enough to the position of previously placed particles. If the particles interact with a Lennard-Jones potential, the repulsive part will be significant. The total energy of interaction will then become very large and the Boltzmann factor very small.

Indeed, most of the phase space of dense systems corresponds to non-physical configurations with strong particle overlaps and very high energies. Effectively with a zero Boltzmann factor, these configurations contribute nothing to the ensemble average of a property.

Metropolis and Ulam determined how to generate these configurations that contribute to the sums in Eq. 15.5. The idea is simple: instead of all microscopic states of phase space being sampled with uniform probability, more probable states are sampled more frequently than less probable ones. This is the importance sampling method, best described by the 1953 paper by Metropolis, Rosenbluth, Rosenbluth, Teller, and Teller (see Further reading). To explain the importance sampling method we first need to introduce Markov chains, which are an important component of the algorithm. We do this next.

15.3 Markov chains in Monte Carlo

Consider a discrete sequence of events

$$E_1, E_2, \ldots, E_n. \tag{15.6}$$

The chain of events is called a Markov chain if the conditional probability of a specific outcome E_{n+1}, provided a sequence of given previous outcomes $E_1, E_2, \ldots, E_n$, is

$$P(E_{n+1}|E_1, E_2, \ldots, E_n) = P(E_{n+1}|E_n). \tag{15.7}$$

This means that for a Markov chain the probability of each outcome only depends on the immediately previous event. We also call the Markov chain a single step transition probability chain.

Consequently, the joint probability of a certain sequence of events is simply

$$P(E_1, E_2, \ldots, E_n) = P(E_2|E_1)P(E_3|E_2)\ldots P(E_n|E_{n-1}). \tag{15.8}$$

Metropolis Monte Carlo algorithms generate a Markov chain of states in phase space. That is, each new state generated is not independent of the previously generated ones. Instead, the new state depends on the immediately preceding state.

The method has one important condition: the outcome of a trial belongs to a finite set of outcomes. In other words, there is a finite number, $\mathcal{N}$, of phase state points $[\underline{X}_1, \underline{X}_2, \ldots, \underline{X}_{\mathcal{N}}]$ corresponding to a macroscopic, equilibrium state.

With a finite number of points, we can define the transition probability between any two state points, $\underline{X}_m$ and $\underline{X}_n$. Denoted with π_{mn}, this is the probability that a point-generating trial produces $\underline{X}_n$ given the previously generated point was $\underline{X}_m$.

The transition matrix $\underline{\underline{\Pi}}$ is the $\mathcal{N} \times \mathcal{N}$ matrix of all transition probabilities between points in the configurational part of the phase space. A key concept is that the sequence of outcomes in a Markov chain is solely governed by the transition matrix. To illustrate this point consider the following example.

Example 15.1

Let us consider the weather in Minneapolis, Minnesota (no Ole and Lena jokes, please). For simplicity's sake let us assume that there are only two possible random outcomes for each day, sunny or cloudy. The state space of stochastic outcomes has then a size of two. Let us denote "sunny" and "cloudy," the two available points in state space, with "S" and "C," respectively.

Let us also assume that a succession of daily weather outcomes forms a Markov chain, and that there exists a specific pattern of weather behavior as follows:

- if it is sunny one day, there is a 70% chance of being sunny the next day. The sunny-to-sunny transition probability is then $\pi_{SS} = 0.70$. Consequently, the sunny-to-cloudy transition probability is $\pi_{SC} = 0.30$.
- if it is cloudy one day, there is a 45% chance of being cloudy the next day. The cloudy-to-cloudy transition probability is $\pi_{CC} = 0.45$. Consequently, the cloudy-to-sunny transition probability is $\pi_{CS} = 0.55$.

The sequence of weather outcomes over many days is a Markov chain, because the outcome for one day depends on the outcome of the previous day.

The transition matrix of this Markov process is then defined as

$$\underline{\underline{\Pi}} = \begin{pmatrix} \pi_{SS} & \pi_{SC} \\ \pi_{CS} & \pi_{CC} \end{pmatrix}$$
$$= \begin{pmatrix} 0.7 & 0.3 \\ 0.45 & 0.55 \end{pmatrix}. \tag{15.9}$$

We can use the transition matrix to find the probability of a sunny or a cloudy day outcome. Let us assume that on day 1 the weather in Minneapolis is equally likely to be sunny or cloudy. We define the probability distribution, ρ, as the two-dimensional vector with the probability of the two states as elements. For the first day, we write

$$\rho(1) = (P_S(1), P_C(1))$$
$$= (0.50, 0.50). \tag{15.10}$$

More generally, $P_S(\tau)$, $P_C(\tau)$ are the probabilities of the τth day being sunny or cloudy, respectively.

In order to determine the probabilities that the weather will be sunny or cloudy the second day, we simply multiply the first day probability distribution with the transition matrix. This yields

$$\begin{aligned}
\rho(2) &= (P_S(2), P_C(2)) \\
&= (P_S(1), P_C(1))\underline{\underline{\Pi}} \\
&= (0.575, 0.425).
\end{aligned} \tag{15.11}$$

Similarly, for the third day

$$\begin{aligned}
\rho(3) &= (P_S(3), P_C(3)) \\
&= (P_S(2), P_C(2))\underline{\underline{\Pi}} \\
&= (P_S(1), P_C(1))\underline{\underline{\Pi}}^2 \\
&= (0.59375, 040625)
\end{aligned} \tag{15.12}$$

and so forth.

Continuing this process for subsequent days, it is observed that the probability distribution ρ quickly reaches a limiting distribution, that is

$$\rho(6) = \rho(7) = \cdots = \rho(\infty) = (0.60, 0.40). \tag{15.13}$$

This limiting distribution is simply the probability distribution of the weather in Minneapolis, which for any day has a probability of 0.60 to be sunny (indeed, Minneapolis has sunny weather for around 60% of the year, and this is true regardless of the temperature).

What is remarkable is that the limiting probability distribution is reached from any arbitrary initial probability distribution, $\rho(1)$. Indeed, the limiting distribution of any Markov chain is completely determined by the transition matrix. Let us see how. The transition matrix in Eq. 15.9 is a stochastic matrix. A matrix $\underline{\underline{\Pi}}$ is called a stochastic matrix if

1. the row elements sum up to 1, i.e., $\sum\limits_{m} \pi_{mn} = 1$;

2. all the elements are positive definite, i.e., $\pi_{mn} \geq 0$, for all pairs (m, n).

Generally, the limiting, or stationary, probability distribution is defined as

$$\rho = \lim_{\tau \to \infty} \rho(0)\underline{\underline{\Pi}}^\tau. \tag{15.14}$$

Obviously, the limiting distribution ρ satisfies the eigenvalue equation

$$\rho = \rho\underline{\underline{\Pi}} \tag{15.15}$$

or

$$\sum_{m} \rho(m)\pi_{mn} = \rho(n). \tag{15.16}$$

In Eq. 15.15, ρ is the eigenvector of the stochastic matrix $\underline{\underline{\Pi}}$ corresponding to an eigenvalue of 1. In other words, the limiting probability distribution, ρ, is completely determined by the stochastic matrix, $\underline{\underline{\Pi}}$, and not influenced by the initial values of ρ.

It is notable that there is an infinite number of transition matrices that satisfy Eq. 15.16. For example, a choice of transition matrix elements that satisfy the so-called microscopic reversibility condition, expressed as

$$\rho(m)\pi_{mn} = \rho(n)\pi_{nm}, \tag{15.17}$$

results in a transition matrix with the limiting probability distribution ρ as the eigenvector corresponding to the largest eigenvalue of 1. In the early 1900s, Oskar Perron and Georg Frobenius proved that Markov chains always have a limiting distribution when they are ergodic. A Markov chain is ergodic if any state can be reached from any other state in non-infinite trials, that is when all the transition probabilities are non-zero. This is an extension of the Perron–Frobenius theorem, which asserts that a stochastic matrix has a unique largest real eigenvalue of 1, and that the corresponding eigenvector is the vector having as elements the limiting probability of state space points. The proof of the Perron–Frobenius theorem is beyond the scope of this text. The interested reader can consult the literature references in Further reading.

15.4 Importance sampling

In statistical mechanics the goal is to generate a chain of microscopic states that sample the limiting, equilibrium probability distribution in phase space. Of course, the transition matrix is immense. For example, for NVE ensembles the transition matrix is an Ω^2 matrix. This is incomprehensibly large. In addition, the matrix elements are unknowable, since it is not possible to determine the transition probability between any two arbitrary points in phase space.

The genius of Metropolis and his colleagues is that they recognized that the limiting distribution can be sampled by devising any one of an infinite number of different transition matrices, $\underline{\underline{\Pi}}_{MC}$, that have the limiting distribution as their larger eigenvector, which for any stochastic matrix corresponds to an eigenvalue of 1.

Proving this is beyond the scope of this text but we can explain it by focusing on the NVT ensemble. The goal of Metropolis Monte Carlo is to generate a Markov chain of configurations

$$\underline{X}_1, \underline{X}_2, \ldots \underline{X}_\lambda, \ldots \underline{X}_{N_{\text{trials}}}$$

using a transition matrix that satisfies the eigenvalue equation

$$\rho = \rho \underline{\underline{\Pi}}_{MC}. \tag{15.18}$$

In Eq. 15.18, ρ is the limiting, equilibrium distribution, with elements

$$\begin{aligned} \rho_\lambda &= \rho_{NVT}(\underline{X}_\lambda) \\ &= \frac{\exp(-\beta U(\underline{X}_\lambda))}{Z(N, V, T)}. \end{aligned} \tag{15.19}$$

The transition matrix in Metropolis Monte Carlo is constructed so that its largest eigenvector, corresponding to an eigenvalue of 1, is a vector with elements $\rho(\lambda) = \rho_{NVT}(\underline{X}_\lambda)$ for each generated point $\underline{X}_\lambda$ in phase space. This can be further clarified by examining the implementation steps in importance sampling, as the Metropolis Monte Carlo algorithm is also known.

There are two major steps in computational implementation of importance sampling:
1. generation of a point in phase space using information of the previously generated phase point;
2. acceptance or rejection of the newly generated point.

The transition matrix is defined through the generation, or attempt, and the acceptance probabilities. In the following sections we present these two important steps.

15.4.1 How to generate states

In principle, we can generate microscopic states uniformly in state space and then apply the Metropolis criterion, as discussed in the next section, to keep states with frequency proportional to their probability. Indeed, states can be generated in any arbitrary way. Of course, for dense systems, unless the states are carefully constructed, the majority of generated states will not be accepted in the growing Markov chain because of their high energy.

In order to generate states that will have a high acceptance probability, in the Metropolis Monte Carlo algorithm an attempt matrix is defined with each element δ_{mn} representing the attempt probability of moving to state $\underline{X}_n$ from state $\underline{X}_m$. In the original algorithm Metropolis and coworkers developed, the attempts are not biased and the attempt matrix is symmetric, i.e., $\delta_{mn} = \delta_{nm}$.

Practically the attempt probabilities are determined by the types of "move" in the program. Bold moves can be attempted that generate microscopic states that are substantially different from previous ones. For example, all particles in the system can be moved by arbitrarily long distances, in arbitrary directions. Less bold moves generate microscopic states close to the previous ones, with attempts made over only small distances of state space. For example, a single particle can be moved by a fraction of its diameter.

The types of attempted move are empirically determined and are system-specific. They are chosen so that a balance is maintained between quickly sampling the phase space and increasing the percentage of accepted states to include in the Markov chain.

Practically, for NVT systems of simple spherical particles, a local environment is defined for each atom, usually a cube of edge R centered at the center of mass of the atom. The number of possible moves N_R in R^3 can be defined by a lattice grid of N_R sites. A new microscopic state can be generated with the move in phase space corresponding to random Cartesian displacements in R^3.

The attempt probability can be defined as

$$\delta_{mn} = \begin{cases} \dfrac{1}{N_R}, & \text{if } \underline{X}_n \text{ is in } R^3 \\ 0. & \text{if } \underline{X}_n \text{ is outside of } R \end{cases} \tag{15.20}$$

More practically, in an algorithm an adjustable parameter $dr_{\max}$ is utilized that controls the boldness of moves. Typically, $dr_{\max}$ is chosen so that 40–50% of the moves are accepted. We see exactly how later in this chapter, when we present a Monte Carlo method pseudo-code.

15.4.2 How to accept states

Assume that the program starts with a point $\underline{X}_m$ and then generates a different point $\underline{X}_n$ in phase space. The algorithm will accept the new point, including it in the chain, with the following acceptance probability:

$$a_{mn} = \begin{cases} 1, & \text{if } \rho_n \geq \rho_m \\ \dfrac{\rho_n}{\rho_m}. & \text{if } \rho_n < \rho_m \end{cases} \tag{15.21}$$

We write more concisely

$$a_{mn} = \min\left(1, \frac{\rho_n}{\rho_m}\right). \tag{15.22}$$

This means that the Metropolis method proceeds from $\underline{X}_m$ to $\underline{X}_n$ with an acceptance probability $\min(1, \rho_n/\rho_m)$. If the new point has a higher probability than the previous one, then the algorithm accepts it and includes it in the chain of outcomes. If the new point has a lower probability the algorithm accepts it with a probability equal to ρ_n/ρ_m.

The Metropolis transition probability is defined with the help of the attempt and acceptance probabilities as follows:

$$\pi_{mn} = \delta_{mn} a_{mn}, \quad m \neq n \tag{15.23}$$

and

$$\pi_{mm} = 1 - \sum_{m \neq n} \pi_{mn}. \tag{15.24}$$

For an NVT system the relative frequency of each configuration is proportional to the corresponding Boltzmann factor. The Metropolis transition probability becomes, assuming $\delta_{mn} = \delta_{nm}$,

$$\pi_{mn} = \min \left(1, \frac{\dfrac{\exp(-\beta U(\underline{X}_m))}{Z(N, V, T)}}{\dfrac{\exp(-\beta U(\underline{X}_n))}{Z(N, V, T)}} \right). \tag{15.25}$$

The configurational integral, $Z(N, V, T)$, cancels out and the transition matrix elements are

$$\pi_{mn} = \min(1, \exp(-\beta \Delta U(\underline{X}))), \tag{15.26}$$

where

$$\Delta U(\underline{X}) = U(\underline{X}_n) - U(\underline{X}_m). \tag{15.27}$$

If the energy of the new state is lower than the energy of the previous state, the algorithm accepts the move in phase space. If the energy is higher the move is accepted with a probability equal to $\exp(-\beta \Delta U(\underline{X}))$.

With this simple acceptance criterion, the Metropolis Monte Carlo method generates a Markov chain of states or conformations that asymptotically sample the NVT probability density function. It is a Markov chain because the acceptance of each new state depends only on the previous state. Importantly, with transition probabilities defined by Eqs. 15.23 and 15.24, the transition matrix has the limiting, equilibrium distribution as the eigenvector corresponding to the largest eigenvalue of 1.

The ensemble average of a macroscopic property $M = M(\underline{X})$ is then

$$\langle M \rangle_{NVT} = \frac{1}{N_{\text{trials}}} \sum_{\lambda=1}^{N_{\text{trials}}} M(\underline{X}_\lambda). \tag{15.28}$$

What is remarkable is that there is no need to multiply $M(\underline{X}_\lambda)$ by the probability density of each state, as in Eq. 15.5. The ensemble average is the arithmetic average because all states are generated in proportion to their probability density. Consequently, the major benefit of importance sampling methods is that they do not generate states that contribute nothing to the ensemble average.

Another important decision relates to the number of trials, N_{trials}. Admittedly, there is no precise a-priori determination of the number of Monte Carlo moves necessary to sample the probability density. The answer is attained empirically, with test simulations. Properties must be monitored, such as the energy and the pair distribution function. When the average properties converge to stable values, the Monte Carlo simulation may be stopped.

15.4.3 *Metropolis Monte Carlo pseudo-code*

We have now described all the steps in a Metropolis Monte Carlo algorithm. Here are the steps that implement this algorithm:

1. Generate an initial microscopic state $\underline{X}_m = (\underline{r}_{1,m}, \underline{r}_{2,m}, \cdots \underline{r}_{N,m})$.
2. Compute the potential energy $U(\underline{X}_m)$.
3. Determine the Boltzmann factor $\exp(-\beta U(\underline{X}_m))$.
4. Generate a neighboring configuration $\underline{X}_n$ by randomly moving one of the N particles within a cubic region. To do this, draw a random number ζ and execute $x_n = x_m + dr_{\max}(\zeta - 0.5)$. Repeat for y_n and z_n.
5. Compute the potential energy $U(\underline{X}_n)$.
6. Determine the Boltzmann factor $\exp(-\beta U(\underline{X}_n))$.
7. If $U(\underline{X}_n) < U(\underline{X}_m)$ accept the new configuration.
8. If $U(\underline{X}_n) > U(\underline{X}_m)$ accept the new configuration with a probability equal to $\exp(-\beta \Delta U(\underline{X}))$, as follows:
   a) Compare $f = \exp(-\beta \Delta U(\underline{X}))$ with a newly drawn random number ζ;
   b) If $\zeta \leq f$ then accept $\underline{X}_n$;
   c) If $\zeta > f$ then reject $\underline{X}_n$, and keep $\underline{X}_m$.

In summary, Metropolis Monte Carlo biases the generation of configurations toward those that make significant contributions to the integral (importance sampling). It generates states with probability proportional to $\exp(-U(\underline{X})/k_B T)$ and counts them equally, whereas simple uniform sampling methods generate states with equal probability, regardless of their energy, and then assigns to them a weight equal to their Boltzmann factor (see Fig. 15.1).

15.4.4 Importance sampling with a coin and a die

Let us visit the weather pattern example again. Let us devise a simple Metropolis Monte Carlo algorithm to generate a chain of "S" and "C" outcomes with a frequency that is proportional to their stationary distribution. Remember that the stationary probability distribution of the weather in Minneapolis is $\rho = (0.60, 0.40)$, with $P_S = 0.60$ and $P_C = 0.40$. For N_{trials} the Monte Carlo algorithm should generate $0.60N_{\text{trials}}$ "S" outcomes, and $0.40N_{\text{trials}}$ "C" outcomes. The point is that generally in Monte Carlo simulations we do not know the actual transition probabilities. Instead, we need to devise a transition matrix that has the stationary probability distribution as its eigenvector, corresponding to the largest eigenvalue of 1. This is a simple enough example that no computer is necessary. Instead we need only a coin and a die.

With the coin, we can generate a uniform attempt distribution of "S" and "C" outcomes. For example, we can assign "S" to heads and "C" to tails. The attempt probability is then $\delta_{SS} = \delta_{SC} = \delta_{CC} = \delta_{CS} = 0.5$.

With the die, we generate the appropriate acceptance criteria. The transitions from "S" to "S," from "C" to "C," and from "C" to "S" will always be accepted, simply because the probability of each new state is higher than or equal to the probability of the previous state.

Figure 15.1 When $(V_n - V_m \leq 0)$ then the transition from state m to state n is always accepted. Otherwise, the transition is accepted with a probability equal to $\exp(-\beta \Delta V)$.

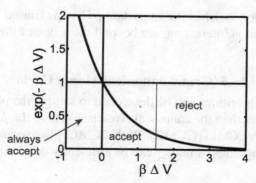

Transitions from "S" to "C" will be accepted with a relative frequency of $P_C/P_S = 0.4/0.6 = 2/3$. We can implement the Metropolis criterion in order to achieve importance sampling by rolling the die: if a "1" or "2" is rolled (probability 1/3), then reject a transition from "S" to "C"; if anything else is rolled (probability 2/3) accept the transition from "S" to "C".

The transition matrix is then, according to Eqs. 15.23 and 15.24:

$$\underline{\underline{\Pi}} = \begin{pmatrix} \pi_{SS} & \pi_{SC} \\ \pi_{CS} & \pi_{CC} \end{pmatrix}$$
$$= \begin{pmatrix} 2/3 & 1/3 \\ 1/2 & 1/2 \end{pmatrix}. \tag{15.29}$$

It is trivial to show that the eigenvector of this stochastic matrix with the largest eigenvalue of 1 is the stationary distribution $\rho = (0.60, 0.40)$.

Note that the transition matrix devised by the Metropolis Monte Carlo method need not be the actual transition matrix, which, for this example, was presented earlier in this chapter. Indeed, simply changing the attempt and acceptance criteria results in an infinite number of matrices that can sample the stationary distribution, only differing in the speed of sampling.

15.4.5 Biased Monte Carlo

We should note that there is no requirement for the attempt matrix to be symmetric. Biased Monte Carlo techniques exist with asymmetric probabilities where the phase space may be sampled more efficiently. One condition, called the microscopic reversibility condition, must be satisfied:

$$\rho_m \delta_{mn} a_{mn} = \rho_n \delta_{nm} a_{nm}. \tag{15.30}$$

This condition satisfies the important $\rho\underline{\underline{\Pi}} = \rho$. Biased Monte Carlo methods, although interesting, are beyond the scope of this text.

15.5 Grand canonical Monte Carlo

Monte Carlo algorithms can be developed to sample the phase space of ensembles other than the canonical. We discuss here the μVT or grand canonical Monte Carlo (GCMC). With GCMC, phase equilibria and the volumetric properties of matter can be efficiently determined.

The limiting distribution of the grand canonical ensemble is given by

$$\rho_{\mu VT}(\underline{X}, N) = \frac{1}{N!\Lambda^{3N}} \frac{\exp(N\beta\mu)\exp(-\beta U(\underline{X}))}{\Xi}. \quad (15.31)$$

GCMC generates a Markov chain of microscopic states that sample this probability distribution. There are two types of move in GCMC:

1. Thermal equilibration moves. These are the same moves in constant NVT phase space with particle displacements and acceptance rules,

$$\min\left(1, \frac{\delta_{nm}}{\delta_{mn}}\exp(-\beta\Delta U)\right). \quad (15.32)$$

2. Chemical equilibration moves. These are particle insertion and deletion moves:

   a. *Particle insertion.* The transition probability of adding a particle from N to $N+1$ particles is

   $$\pi_{mn}^{ins} = \min\left[1, \frac{\delta_{mn}}{\delta_{nm}}\exp\left(-\beta\Delta U + \ln\left(\frac{KV}{N+1}\right)\right)\right], \quad (15.33)$$

   where

   $$K = \exp(\beta\mu)/\Lambda^3 \quad (15.34)$$

   is a constant of the simulation.

   b. *Particle deletion.* The transition probability of deleting a particle from $N+1$ to N particles is

   $$\pi_{mn}^{del} = \min\left[1, \frac{\delta_{mn}}{\delta_{nm}}\exp\left(-\beta\Delta U + \ln\left(\frac{N}{KV}\right)\right)\right]. \quad (15.35)$$

Using these acceptance rules, the system maintains constant μ, V, T.

15.6 Gibbs ensemble Monte Carlo for phase equilibria

Thanos Panagiotopoulos developed an ingenious method to compute phase equilibria of fluids. The basic idea is to simultaneously simulate samples of two bulk phases in equilibrium.

Consider a system in the canonical ensemble. Assume that this system is made up of two subsystems, I and II, of different densities at equilibrium. Consider for example a liquid phase at equilibrium with its vapor. Each of the subsystems is in the μPT ensemble.

For the two systems we have variable numbers of particles, volume, and energy:

$$\begin{cases} \text{System I: } E^I, V^I, N^I, \\ \text{System I: } E^{II}, V^{II}, N^{II}. \end{cases} \quad (15.36)$$

The following constraints apply for the composite system:
1. $N = N^I + N^{II}$,
2. $V = V^I + V^{II}$,
3. $E = E^I + E^{II}$.

Instead of simulating a single system with both coexisting phases, GEMC considers two independent systems, each modeling a single phase. Periodic boundary conditions are used for both systems, effectively modeling two macroscopic phases. Although there is no direct contact between the two phases, the constraints apply to both.

There are three types of GEMC move:
1. *Particle displacements*, in order to equilibrate both systems thermally. These are performed in each system box separately. The same attempt and acceptance rules as before apply.
2. *Volume changes*, in order to achieve mechanical equilibrium. Volume is exchanged between the two subsystems, assuming constant NPT conditions for both. Since the pressure is the same the volume changes are the same but with opposite signs

$$\Delta V^I = -\Delta V^{II} = \Delta V. \quad (15.37)$$

The practical implementation is as follows:
a) Pick a random volume change in box I. The positions of particles can be rescaled to the new volume. Because of the volume change ΔV^I and the change in the particle positions, there is then a change in the potential energy of the system ΔU^I.
b) The ratio of new to old NPT probability densities is

$$P_{\text{vol}}^I = \frac{\rho_{\text{new}}}{\rho_{\text{old}}}, \quad (15.38)$$

where $\rho = \rho_{NPT}$ is the probability density of the NPT ensemble.

Consequently,

$$P_{\text{vol}}^I = \exp[-\beta P \Delta V^I - \beta \Delta U^I + N^I \ln(V^I + \Delta V) - N^I \ln(V^I)]. \quad (15.39)$$

c) Equivalently, the change of volume in box II has a probability

$$P_{vol}^{II} = \exp[-\beta P \Delta V^{II} - \beta \Delta U^{II} + N^{II} \ln(V^I - \Delta V)$$
$$- N^{II} \ln(V^{II})]. \tag{15.40}$$

d) The overall probability of combined moves is

$$P_{vol} = P_{vol}^{I} P_{vol}^{II}. \tag{15.41}$$

e) In the algorithm, the volume change is accepted with

$$\pi_{acc} = \min(1, P_{vol}). \tag{15.42}$$

3. *Particle exchanges*, in order to achieve chemical equilibrium between the phases. We consider two simultaneous grand canonical Monte Carlo moves for the two systems in $\mu V^I T$ and $\mu V^{II} T$ ensembles. The probability of exchanging a particle from II to I is

$$P_{exc} = \exp\left(-\beta \left[\Delta U^I + \Delta U^{II} + k_B T \ln\left(\frac{V^{II}(N^I + 1)}{V^I N^{II}}\right)\right]\right). \tag{15.43}$$

The acceptance criterion is

$$\pi_{acc} = \min(1, P_{exc}). \tag{15.44}$$

The exact sequence of moves is determined empirically. Test runs are typically required to determine an optimum set of parameters for the boldness and the frequency of moves. The interested reader is referred to literature in Further reading, where thorough descriptions of Monte Carlo moves are presented in detail.

15.7 Further reading

1. A. Z. Panagiotopoulos, *Molec. Phys.*, **61**, 813–826, (1987).
2. M. P. Allen and D. J. Tildesley, *Computer Simulation of Liquids*, (London: Oxford University Press, 1989).
3. N. Metropolis, A. W. Rosenbluth, M. N. Rosenbluth, A. H. Teller, and E. Teller, Equation of state calculations by fast computing machines, *J. Chem. Phys.*, **21** (6), 1087–1092, (1953).
4. S. P. Meyn and R. L. Tweedie, *Markov Chains and Stochastic Stability*, (London: Springer-Verlag, 1993).
5. J. M. Hammersley and D. C. Handscomb, *Monte Carlo Methods*, (London: Methuen, 1975).

6. K. Binder, *The Monte Carlo Method in Condensed Matter Physics*, (New York: Springer, 1995).
7. R. Y. Rubinstein and D. P. Kroese, *Simulation and the Monte Carlo Method*, (New York: John Wiley & Sons, 2007).

15.8 Exercises

1. Prove that a matrix whose elements satisfy the microscopic reversibility condition satisfies Eq. 15.16.

2. Prove that for the transition matrix in Metropolis Monte Carlo, constructed with Eq. 15.26, the largest eigenvalue is 1 and the corresponding eigenvector is a vector with elements $\rho(\lambda) = \rho_{NVT}(\lambda)$ for each generated point $\underline{X}_\lambda$ in phase space.

3. Develop an importance sampling algorithm that samples the Maxwell–Boltzmann distribution of velocities of simple particles.

4. Arguably, the book *Computer Simulations of Liquids* by Allen and Tildesley is the *bible* of simulations. They have written an extensive library of FORTRAN codes that can be found in the Computational Chemistry List (http://www.ccl. net/cca/software/SOURCES/FORTRAN/allen-tildesley-book/f.00.shtml).

    Combine subroutines f.23 and f.11 to write a program that creates an FCC configuration of particles, and then reads it and runs an MC simulation. Run two simulations at different reduced densities with 264 particles and report on the properties that each simulation calculates. Suggest other properties that can be calculated by the simulation.

16

Molecular dynamics simulations

Molecular dynamics (MD) simulations generate trajectories of microstates in phase space. They simulate the time evolution of classical systems by numerically integrating the equations of motion.

With their strength tied to available computer speed, simulations continue to become a more powerful tool. A letter to the *Journal of Chemical Physics* by B. J. Alder and T. E. Wainwright in 1957 was the first work that reported results from molecular dynamics simulations. The Lawrence Radiation Laboratory scientists studied two different sized systems with 32 and 108 hard spheres. They modeled bulk fluid phases using periodic boundary conditions. In the paper they mention that they counted 7000 and 2000 particle collisions for 32 and 108 particle systems, respectively. This required one hour on a UNIVAC computer. Incidentally, this was the fifth such commercial computer delivered out of the 46 ever produced. The computer cost around $200 000 in 1952 and each of ten memory units held 100 words or bytes. Nowadays, a $300 personal computer with a memory of approximately 500 000 000 bytes can complete this simulation in less than 1 second. And Moore's empirical law that computer power doubles every 18 months still holds.

Alder and Wainwright computed the pressure of the system and their results compared favorably to results of Monte Carlo simulations of the same system. In the paper, they state in an unduly reserved manner that "this agreement provides an interesting confirmation of the postulates of statistical mechanics for this system." Indeed the ergodic hypothesis has been validated time and again over the last five decades with results matching from both Monte Carlo and molecular dynamics simulations for a terrific variety of systems.

An advantage of molecular dynamics simulations is that time is a simulation variable. Positions and velocities of particles are calculated as functions of time. With these, dynamic properties, such as the diffusion coefficient, can be determined. We discuss calculation of properties from simulation results in the following chapter. Herein, we describe the algorithms for conducting molecular dynamics simulations.

16.1 Molecular dynamics simulation of simple fluids

Let us consider a system of N monoatomic particles interacting with Lennard-Jones interactions. The molecular model construction has been described in Chapter 14.

The classical equations of motion for particles of mass m are $3N$ second-order differential equations

$$m\underline{\ddot{r}}_i = \underline{F}_i, \tag{16.1}$$

where the force on particle i is calculated from the potential energy as follows:

$$\underline{F}_i = -\underline{\nabla}U. \tag{16.2}$$

Equivalently, we can write the following $6N$ first-order differential equations

$$\underline{\dot{r}}_i = \frac{\underline{\dot{p}}_i}{m_i},$$
$$\underline{\dot{p}}_i = \underline{F}_i. \tag{16.3}$$

In practice, numerical integration of $6N$ first-order differential equations is less challenging and computationally demanding than numerical integration of $3N$ second-order differential equations. Thus, the set of Eq. 16.3 is the preferred one for MD simulations.

We should note that in the absence of any external fields, the total energy of the system is conserved. In other words, numerical integration of Newton's equations of motion simulates the microcanonical ensemble. The forces are then called conservative.

The equations of motion can be altered also to simulate other ensembles. We discuss methods for simulating canonical and isothermal-isobaric ensembles later in this chapter.

16.2 Numerical integration algorithms

In Chapter 14 we discussed the generation of an initial configuration, $(\underline{r}(0), \underline{v}(0))$. The equations of motion can then be integrated numerically, with this phase space point as the initial condition.

The integration can be implemented with a variety of finite difference algorithms. These propagate the positions, velocities, accelerations, third derivative of positions with respect to time, and so on, at a later time $t + \delta t$, given the values of these variables at time t. The time interval δt is the numerical integration time step.

Typically, we assume that positions and their time derivatives can be written as Taylor series expansions. The positions are written as

$$\underline{r}(t + \delta t) = \underline{r}(t) + \delta t \, \underline{v}(t) + \frac{1}{2} \delta t^2 \underline{a}(t) + \frac{1}{6} \delta t^3 \underline{b}(t) + \dots \quad (16.4)$$

Similarly, the velocities are written as

$$\underline{v}(t + \delta t) = \underline{v}(t) + \delta t \, \underline{a}(t) + \frac{1}{2} \delta t^2 \underline{b}(t) + \frac{1}{6} \delta t^3 \underline{c}(t) + \dots \quad (16.5)$$

Analogous expansions can be written for the accelerations, $\underline{a}(t)$, the third derivative of the positions with respect to time, $\underline{b}(t)$, the fourth derivative of the positions with respect to time, $\underline{c}(t)$, and so on.

Various finite difference algorithms exist, identified by their truncation schemes. For example, Euler's method* includes only the positions and velocities, ignoring the third and higher order derivatives:

$$\underline{r}(t + \delta t) = \underline{r}(t) + \delta t \, \underline{v}(t) \quad (16.6)$$

and

$$\underline{v}(t + \delta t) = \underline{v}(t) + \delta t \, \underline{a}(t). \quad (16.7)$$

The choice of method depends on the acceptable amount of error generated by the calculation. There are two important types of numerical error in any finite differences algorithm:
1. *Truncation errors.* These are the result of the way the integration method approximates the true solution of the equations of motion. The error is proportional to the largest term ignored in the expansion. For example, the truncation error in the Euler method is of order $O(\delta t^2)$. An algorithm is called nth order if the truncation error is of order $O(\delta t^{n+1})$. The Euler algorithm is a first-order method.
2. *Roundoff errors.* These are associated with the way the computer represents real variables. Only a finite number of decimal places is retained by a computer in its random access memory and the

---

* Leonhard Euler (1707–1783) was not interested in numerical integration of ordinary differential equations, but he was interested in initial value problems. Incidentally, Euler is also responsible for what, according to Richard Feynman, is the most remarkable equation ever: $e^{i\pi} + 1 = 0$.

computer truncates all higher places. Roundoff errors are machine and compiler specific. The reader who is interested in learning more about how modern computers represent real numbers with floating point representations may consult Wikipedia for a quick reference (search "floating point" at www.wikipedia.org).

Every care must be taken when developing an algorithm or conducting a simulation to minimize these errors. Numerical errors not only impact the accuracy of the simulation results, they also affect the numerical stability of the algorithm. Accumulation of numerical errors may result in numerical overflow (e.g., when a variable becomes larger than the maximum value a computer can handle) and a program crash.

As anticipated, the higher the order of the algorithm, the higher the accuracy. Yet, higher accuracy comes at increasing computational cost. There are many algorithms that nicely balance the demands for accuracy, stability, and computational efficiency. We present two large classes, widely used by molecular simulations specialists, the predictor–corrector algorithms and the Verlet algorithms.

16.2.1 Predictor–corrector algorithms

In 1971, Bill Gear developed what is now called the four-value predictor–corrector algorithm. There are two steps for propagating the positions, velocities, accelerations, and third derivatives of positions. First is the prediction step, which is implemented as follows:

$$\underline{r}_p(t + \delta t) = \underline{r}(t) + \delta t \underline{v}(t) + \frac{1}{2}\delta t^2 \underline{a}(t) + \frac{1}{6}\delta t^3 \underline{b}(t), \qquad (16.8)$$

$$\underline{v}_p(t + \delta t) = \underline{v}(t) + \delta t \underline{a}(t) + \frac{1}{2}\delta t^2 \underline{b}(t), \qquad (16.9)$$

$$\underline{a}_p(t + \delta t) = \underline{a}(t) + \delta t \underline{b}(t), \qquad (16.10)$$

$$\underline{b}_p(t + \delta t) = \underline{b}(t). \qquad (16.11)$$

Note that the equations of motion have yet to be used. They are implemented in the corrector step.

The forces on all particles $\underline{f}(t + \delta t)$ are computed using the positions $\underline{r}_p(t + \delta t)$. The corrected accelerations are then determined as follows:

$$\underline{a}_c(t + dt) = \underline{f}(t + \delta t)/m. \qquad (16.12)$$

Defining

$$\Delta \underline{a}(t + \delta t) = a_c(t + \delta t) - a_p(t + \delta t), \qquad (16.13)$$

the positions and velocities are corrected as follows:

$$\underline{r}_c(t + dt) = \underline{r}_p(t + \delta t) + C_1 \Delta \underline{a}(t + \delta t) \tag{16.14}$$

and

$$\underline{v}_c(t + dt) = \underline{v}_p(t + \delta t) + C_2 \Delta \underline{a}(t + \delta t). \tag{16.15}$$

The accelerations and third position derivatives are also corrected:

$$\underline{a}_c(t + dt) = \underline{a}_p(t + \delta t) + C_3 \Delta \underline{a}(t + \delta t) \tag{16.16}$$

and

$$\underline{b}_c(t + dt) = \underline{b}_p(t + \delta t) + C_4 \Delta \underline{a}(t + \delta t). \tag{16.17}$$

The constants depend on the order of the integration scheme. For a fourth-order predictor corrector scheme, Allen and Tildesley calculate the constants to have the following values:

$$C_1 = \frac{1}{6}, \quad C_2 = \frac{5}{6}, \quad C_3 = 1, \quad C_4 = \frac{1}{3}. \tag{16.18}$$

16.2.2 *Verlet algorithms*

Another widely used class of algorithms is the Verlet algorithms. These were first developed by Loup Verlet in 1967. Herein, we present three variants.

1. Simple Verlet algorithm.

The algorithm starts with

$$\underline{r}(t + \delta t) = \underline{r}(t) + \delta t \underline{v}(t) + \frac{1}{2} \delta t^2 \underline{a}(t) + \cdots \tag{16.19}$$

and

$$\underline{r}(t - \delta t) = \underline{r}(t) - \delta t \underline{v}(t) + \frac{1}{2} \delta t^2 \underline{a}(t) - \cdots \tag{16.20}$$

Combining Eqs. 16.18 and 16.20 yields

$$\underline{r}(t + \delta t) = 2\underline{r}(t) - \underline{r}(t - \delta t) + \delta t^2 \underline{a}(t). \tag{16.21}$$

The velocities do not enter into the simple Verlet algorithms. They are subsequently determined in the following separate step:

$$\underline{v}(t) = \frac{(\underline{r}(t + \delta t) - \underline{r}(t - \delta t))}{2\delta t}. \tag{16.22}$$

2. Leapfrog Verlet algorithm.

In this algorithm the positions and velocities are determined as follows:

$$r(t + \delta t) = r(t) + \delta t \underline{v}\left(t + \frac{1}{2}\delta t\right) \tag{16.23}$$

and

$$\underline{v}\left(t + \frac{1}{2}\delta t\right) = \underline{v}\left(t - \frac{1}{2}\delta t\right) + \delta t \underline{a}(t). \tag{16.24}$$

For $\underline{v}(t)$ we compute

$$\underline{v}(t) = \frac{1}{2}\left(\underline{v}\left(t + \frac{1}{2}\delta t\right) - \underline{v}\left(t - \frac{1}{2}\delta t\right)\right). \tag{16.25}$$

3. Velocity Verlet algorithm.

The positions and velocities are calculated as follows:

$$r(t + \delta t) = r(t) + \delta t \underline{v}(t) + \frac{1}{2}\delta t^2 \underline{a}(t) \tag{16.26}$$

and

$$\underline{v}(t + \delta t) = \underline{v}(t) + \frac{1}{2}\delta t \left(\underline{a}(t) + \underline{a}(t + \delta t)\right). \tag{16.27}$$

The specific steps for the velocity Verlet algorithm are the following:

a) Determine $r(t + \delta t)$ with Eq. 16.26.

b) Compute the forces $\underline{f}(t + \delta t)$ and accelerations $\underline{a}(t + \delta t)$ with $r(t + \delta t)$.

c) Calculate the velocities with the following:

$$\underline{v}\left(t + \frac{1}{2}\delta t\right) = \underline{v}(t) + \frac{1}{2}\delta t \underline{a}(t), \tag{16.28}$$

and

$$\underline{v}(t + \delta t) = \underline{v}\left(t + \frac{1}{2}\delta t\right) + \frac{1}{2}\delta t \underline{a}(t + \delta t). \tag{16.29}$$

An attractive feature of the Verlet algorithms is that they can be nicely used for algorithms with constraints. In these algorithms, degrees of freedom may be constrained in order to gain computational efficiency. For example, the length of molecular bonds may be constrained, fixing the distance between two bonded atoms. Additional terms need to be included in the equations of motion for the constraints. Verlet algorithms handle these terms quickly and accurately. The interested reader may consult the references at the end of this chapter.

16.3 Selecting the size of the time step

A delicate balance must be struck in the choice of the integration time step size (Fig. 16.1). Too small a time step and adequate exploration and sampling of phase space take inordinate amounts of time. Too big a time step and integration errors lead to inaccuracies and numerical instabilities.

An empirical and useful rule of thumb considers the velocities of particles. In particular, the time step is chosen based on how particle distances change in comparison to the interatomic potential range. For example, argon atoms at room temperature in a low-density state have a most probable velocity of $353\,\text{ms}^{-1}$. With this speed, a time step of $10^{-12}\,\text{s}$ would result in two particles colliding (placed on top of one another) in the next time step, if they were originally at a distance of 2.5σ (there are no interactions at this distance) and moving in opposite

Figure 16.1 The choice of the integration time step δt is important. If the time step is too small the distances particles travel may be too short (from configuration A to configuration B), requiring inordinate amounts of computational time). If the time step is too big, the distances particles travel may be too long (from configuration A to configuration C), resulting in numerical instabilities.

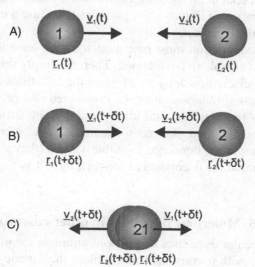

directions toward one another. The energy would then increase from zero at the previous step to incalculable in the current step and the program would crash ("crash" may be too strong a word for a program simply stopping; nonetheless, the loss of simulation results may indeed be painful). On the other hand, a time step of 10^{-18} s would result in thousands of unnecessary integration rounds before the two atoms "see" each other.

If molecular systems are simulated, the time step selection need also consider other degrees of freedom, besides translational ones. The bond, bond-angle, and dihedral-angle vibrations dictate the size of the integration time step. The following time steps are proven empirically to result in stable numerical simulations:

	Type of motion	Suggested time step
Atoms	Translation	10^{-14} s or 10 fs
Rigid molecules	Rotation	5×10^{-15} s or 5 fs
Flexible molecules	Torsion	2×10^{-15} s or 2 fs

16.4 How long to run the simulation?

First, since the initial configuration is constructed arbitrarily, simulations must be conducted long enough for the system to lose memory of the initial configuration. This means that simulated equilibrium properties must not be dependent on the initial condition. A simple empirical rule for the equilibration of liquids is for particles to diffuse a distance equal to two times their diameter.

Second, the simulation must be conducted long enough for an adequate statistical sample to be obtained. There is simply no way to know beforehand precisely how long to integrate the equations of motion for any system. Test simulations must be conducted and properties monitored, such as the potential and kinetic energy, the structure and the pair distribution function, the mean squared displacement, etc. When the average properties converge to stable values (they do not change even if the simulation is conducted longer), then it is time to stop the simulation.

16.5 Molecular dynamics in other ensembles

Standard molecular dynamics simulations simulate the microcanonical ensemble and, with a constant Hamiltonian, the kinetic and potential energies fluctuate with opposite signs (Fig. 16.2).

If the kinetic energy is $K(t)$ at time t, the instantaneous temperature is

$$T(t) = \frac{2K(t)}{3Nk_B},$$ (16.30)

and the average temperature is determined as

$$<T> = \frac{1}{t_{sim}} \sum_{t=1}^{t_{sim}} \frac{2K(t)}{3Nk_B},$$ (16.31)

where t_{sim} is the total simulated number of steps.

The fluctuations in the kinetic energy are found to be

$$<\delta K^2> = \frac{3}{2}Nk_B^2 <T>^2 \left(1 - \frac{3Nk_B}{2C_V}\right).$$ (16.32)

The pressure also fluctuates. The instantaneous values can be determined with the following expression:

$$P = \rho k_B T + \frac{1}{3V} \sum_{i>j} (\underline{r}_i - \underline{r}_j)\underline{f}_{ij},$$ (16.33)

where $\underline{f}_{ij}$ is the force on particle i due to particle j.

In general, if we want to calculate NVT or NPT ensemble properties, there is no simple way to conduct NVE simulations and change the energy E and volume V until the temperature and pressure of the system reach desired values.

Figure 16.2 Solving the equations of motion samples the microcanonical ensemble. The potential and kinetic energies fluctuate in opposite directions so that the total energy remains constant. In this figure the results are from a molecular dynamics simulation of a simple Lennard-Jones fluid.

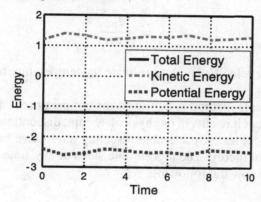

We can, however, modify the equations of motion and simulate in the appropriate ensemble. In the following sections, we present some of the most widely used methods for simulating systems at constant temperature and pressure.

16.5.1 Canonical ensemble molecular dynamics simulations

We present two major classes of NVT algorithms. These are widely used methods in molecular dynamics simulations, but are by no means the only ones. The interested reader is referred to a number of excellent textbooks in Further reading that detail simulation methodologies.

Rescaling methods

These methods continuously scale the velocities so that the desired temperature T_{bath} is attained. The algorithmic steps at any time t during the simulation are as follows:

1. Calculate the current temperature $T(t)$ with Eq. 16.30.
2. Determine the difference between the actual temperature and the targeted one,

$$\Delta T = T_{bath} - T(t). \tag{16.34}$$

3. Define a rescaling constant λ, such that

$$\Delta T = \frac{1}{2} \sum_{i=1}^{N} \frac{2}{3} \frac{m(\lambda v_i)^2}{N k_B} - \frac{1}{2} \sum_{i=1}^{N} \frac{2}{3} \frac{m v_i^2}{N k_B}, \tag{16.35}$$

where v_i is the velocity of particle i.
4. Combine Eq. 16.34 with Eq. 16.35 and solve for λ to find

$$\lambda = \sqrt{\frac{T_{bath}}{T(t)}}. \tag{16.36}$$

5. Multiply all velocities by λ to re-scale the temperature to the correct value.

This is a simple method that ensures the temperature is always at the correct value, but it results in unphysical, abrupt, discontinuous changes in particle velocities. A preferable rescaling alternative borrows from systems control theory. The algorithmic steps at any time t during the simulation are now the following:

1. Calculate the current temperature $T(t)$ with Eq. 16.30.
2. Determine a scaled difference between the actual temperature and the targeted one,

$$\Delta T = \frac{\delta t}{\tau} (T_{\text{bath}} - T(t)), \qquad (16.37)$$

where δt is the integration time step, and τ is a user-defined constant that couples the system to the heat bath. Its value determines how tightly the two are coupled together. If it is large then the coupling is weak. Usually if $\delta t = 10^{-15}$ s then $\tau = 0.4 \times 10^{-12}$ s.
3. Define λ as in Eq. 16.35.
4. Solve for λ to find

$$\lambda = \sqrt{\left(1 + \frac{\delta t}{\tau} (T_{\text{bath}} - T(t))\right)}. \qquad (16.38)$$

5. Multiply all velocities by λ to re-scale the temperature to the correct value.

This rescaling method does not result in velocity discontinuities. Nonetheless, the sampling of the NVT ensemble is still artificial.

Extended NVT methods: the Nose–Hoover method

Ingenious work by Shuichi Nose (1987) and William Hoover (1985) resulted in an algorithm that rigorously samples the NVT ensemble. They introduced a degree of freedom, s, to represent the temperature reservoir, or heat bath, with which the system exchanges energy to maintain a constant temperature. This additional degree of freedom extends the simulated system. It has an associated mass Q_s, momentum p_s, and both a kinetic energy K_s, and a potential energy U_s.

The Hamiltonian of the extended system is then

$$H_E = K + U + K_s + U_s. \qquad (16.39)$$

The total Hamiltonian, H_E, is considered constant, equal to E_E.

The extended system is in the microcanonical ensemble with a probability density

$$\rho_{NVE}(\underline{r}, \underline{p}, s, p_s) = \frac{\delta(H_E - E_E)}{\int d\underline{r} \, d\underline{p} \, ds \, dp_s \delta(H_E - E_E)}. \qquad (16.40)$$

Integration over the heat bath degrees of freedom, s and p_s, yields a canonical distribution for the $(\underline{r}, \underline{p})$ part of the phase space, which pertains to the system. This is expressed as

$$\rho_{NVT}(\underline{r}, \underline{p}) = \frac{\exp(-(K + U)/k_B T)}{\displaystyle\int d\underline{r}\,d\underline{p}\,\exp(-(K + U)/k_B T)}. \tag{16.41}$$

This result is attainable if the potential energy V_s is defined as

$$V_s = (f + 1)\ln(s k_B T), \tag{16.42}$$

where f is the number of degrees of freedom of the system.

The kinetic energy of the bath degree of freedom is defined as

$$K_s = \frac{1}{2} Q_s \dot{s}^2 = \frac{p_s^2}{2 Q_s}. \tag{16.43}$$

The parameter Q_s can be thought of as a thermal inertia that controls the rate of temperature fluctuations inside the system. It is an adjustable parameter that can be empirically determined. If it is too high then the energy transfer between the system and the bath is slow, with the limit of $Q_s \to \infty$ returning the NVE ensemble. On the other hand, if Q_s is too small then there will be long-lived, weakly damped oscillations of the energy in the system.

Nose and Hoover determined that the velocities of the particles need to be rescaled by s, so that $v(t) = sv(t)$ at every time step. The value of s changes according to an additional set of equations of motion. The equations of motion for the extended system are then determined to be

$$\ddot{\underline{r}} = \frac{f}{ms^2} - 2\frac{\dot{s}\dot{r}}{s} \tag{16.44}$$

and

$$Q\ddot{s} = \sum_i m\dot{r}_i s - \frac{(f + 1)k_B T}{s}. \tag{16.45}$$

The derivation of these relations is lucidly described in the papers published by Nose and Hoover in the mid-1980s (see Further reading).

16.6 Constrained and multiple time step dynamics

In complex molecular systems, there are motions with disparate characteristic time scales: fast degrees of freedom, such as bond and angle vibrations, and slow degrees of freedom, such as molecular translation and rotation. The integration time step size is determined by fast degrees of freedom, and is typically chosen to be smaller than their character-istic times. For example, when simulating water, the time step must be smaller than the inverse of the bond vibration frequency, which depends on the temperature. A larger time step would result in inaccurate sam-pling of the bond length.

If the characteristic time scales of the important motions are spread over a wide spectrum, the system is called stiff. Because of the require-ment of small time steps, long simulation times are computationally hard to reach.

Over the last three decades, scientists have proposed a number of ingenious algorithmic solutions for stiff molecular systems. Two are worth mentioning here (albeit briefly, since presentation of the details is beyond the purview of this book):

1. *Constraint dynamics.*

   The bond length vibrations are fast and on occasion unimportant for calculation of accurate structural and thermodynamic properties. They can then be constrained, fixing the atomic distances at the equilibrium value. Bond angle vibrations can also be constrained, using pseudo-bonds to keep triplets of atoms at correct angles.

   In algorithms, each constraint is translated in a force acting along the bond or pseudo-bond. These forces are taken into account in the equations of motion with additional terms. Perhaps the two most widely used algorithmic implementations of constraints are the SHAKE (developed by Ryckaert, Ciccotti, and Berendsen in 1977) and RATTLE (developed by Andersen in 1983) algorithms (see Further reading).

2. *Multiple time step dynamics.*

   In multiple time step algorithms, forces are classified into groups according to how rapidly they change with time. Each group is assigned a separate integration time step and the integration proceeds in loops over the different force groups while maintaining accuracy and numerical stability. In 1992, Tuckerman, Berne, and Martyna developed what is now the most widely used multiple time step algo-rithm, called r-RESPA, which stands for reversible reference system propagation algorithm.

16.7 Further reading

1. M. P. Allen and D. J. Tildesley, *Computer Simulation of Liquids*, (London: Oxford University Press, 1989).

2. D. Frenkel and B. Smit, *Understanding Molecular Simulation: from Algorithms to Applications*, (San Diego, CA: Academic Press, 2002).
3. C. W. Gear, *Numerical Initial Value Problems in Ordinary Differential Equations*, (New York: Prentice-Hall, 1971).
4. L. Verlet, *Phys. Rev.*, **159**, 98, (1967).
5. S. Nose, *J. Chem. Phys.*, **81**, 511–519, (1984).
6. W. G. Hoover, *Phys. Rev.*, A **31**, 1695–1697, (1985).
7. J.-P. Ryckaert, G. Ciccotti, and H. J. C. Berendsen, *J. Comp. Phys.*, **23**, 327–341, (1977).
8. H. C. Andersen, *J. Comp. Phys.*, **52**, 24–34, (1983).
9. M. E. Tuckerman, B. J. Berne, and G. J. Martyna, *J. Chem. Phys.*, **97**, 1990–2001, (1992).

16.8 Exercises

1. Prove Eq. 16.32.

2. Derive Eq. 16.41 from Eq. 16.40, using Eqs. 16.44 and 16.45.

3. Obtain Moldyn from the textbook website (http://statthermo.sourceforge.net/). Moldyn is a set of codes which simulates point mass particles that interact with Lennard-Jones forces. Moldyn uses the velocity–Verlet integration algorithm in the *NVE* ensemble. Read the README file on how to run the programs and do this homework problem.

   Run multiple simulations at different reduced densities (start at $\rho^* = 0.1$ and increase to $\rho^* = 0.8$). Then do the following:
   a) Plot the pair distribution function at different densities.
   b) Calculate the diffusion coefficient at different densities.
   c) Calculate the pressure at the two different simulations (hint: calculate the virial coefficient during the simulation).
   Discuss the results.

   Assume the simulated particles are a) argon, b) methane, c) oxygen. What actual density conditions are simulated?

   Compare thermodynamic properties to experimental values (you can find these in the NIST Chemistry WebBook, at http://webbook.nist.gov/chemistry/).

   You can view a movie of the simulation running JMOL and reading the file animate.xyz produced by the MD simulations you have run. JMOL is also available on sourceforge at http://jmol.sourceforge.net/.

4. Compute and plot the pair distribution function $g(r)$ from an MD simulation of Lennard-Jones particles up to a reduced distance $r^* = 2.5$ at $T^* = 1.084$ and $\rho^* = 0.805$. Compare your simulation results to literature results.

17

Properties of matter from simulation results

The output of simulations is a list of microscopic states in phase space. These are either a sample of points generated with Metropolis Monte Carlo, or a succession of points in time generated by molecular dynamics. Particle velocities, as well as higher time derivatives of particle positions, can also be saved during a molecular dynamics simulation.

This information is periodically recorded on computer disks during the simulation for subsequent analysis. Minimal analysis can be conducted during the actual simulation, primarily in order to ensure that equilibrium is reached and the simulation is numerically stable. Usually most of the analysis is conducted off-line, after the simulation is completed. The drawback is that large files must be stored and processed. Furthermore, saving positions and velocities does not occur for every integration time step. Instead, positions and velocities are saved infrequently, which results in some information being thrown away. It is more convenient, however, to analyze results off-line, developing and using analysis algorithms as needed. It is also important to maintain an archival record of simulation results.

17.1 Structural properties

The pair distribution function, $g(r)$, can be computed from stored particle positions. A program to compute it is given below and can be found in http://statthermo.sourceforge.net/. It computes the distance between all pairs of particles, and counts the number of particle pairs that are in a specific distance range, defined by a parameter *delr*. The program uses these numbers to generate a histogram, which when properly normalized yields the pair distribution function. The number of histogram bins is also a user-defined parameter, *maxbin*.

```
c Zero out the histogram values.
do bin=1, maxbin
hist(bin)=0
enddo

c Open loop for all time points that positions were saved at.
do t=1, trun

c Loop over atoms
do i=1, natom-1

c Use auxiliary variables for computer speed.
rxi=rx(t,i)
ryi=ry(t,i)
rzi=rz(t,i)

c Loop over other atoms
do j=i+1, natom

c Determine distance between i and j
drx=rxi-rx(t,j)
dry=ryi-ry(t,j)
drz=rzi-rz(t,j)

c Apply periodic boundary conditions.
drx=drx-lsimbox*anint(drx*lsimbi)
dry=dry-lsimbox*anint(dry*lsimbi)
drz=drz-lsimbox*anint(drz*lsimbi)
rijsq=drx*drx+dry*dry+drz*drz
rij=sqrt(rijsq)

c Determine the bin number to add this pair in.
bin=int(rij/delr) + 1

if (bin.le.maxbin) then

c Add 2 to the histogram at the specific bin. One for ij and
  another for ji.
hist(bin)=hist(bin) + 2

endif

c Close loops
enddo
enddo
enddo
```

```
c Determine a normalization constant, using the known system
  density, dens
const = 4.0d0 * 3.1415926536 * dens / 3.0d0

do bin=1, maxbin

c Determine pdf, gr, as a function of distances. dfloat(x)
  converts integer x to
c a double precision real number.

rlower=dfloat(bin-1)*delr
rupper=rlower+delr
nideal=const*(rupper**3-rlower**3)
gr(bin)=dfloat(hist(bin))/dfloat(trun*natom)/nideal

c Write distances and pair distribution function in file,
  previously opened with #25.
write(25,*) rlower+delr/2.0d0, gr(bin)

enddo
return
end
```

With the pair distribution function at hand, we can compute thermo-dynamic properties, starting with the excess energy and the pressure as discussed in Chapter 9.

17.2 Dynamical information

17.2.1 Diffusion coefficient

The diffusion coefficient can be readily computed from the mean squared displacement of particles using the following expression:

$$D = \lim_{t \to \infty} \frac{1}{6t} \langle |\underline{r}(t) - \underline{r}(0)|^2 \rangle. \tag{17.1}$$

In practice, the squared displacement of each particle is computed, the displacements of all particles are added and then divided by N to determine the mean. The diffusion coefficient can be calculated from the slope of the mean square displacement as a function of time.

It is important to note that the diffusion coefficient is not accu-rately calculated when using the positions of particles after applying periodic boundary conditions. Periodic boundary conditions constrain the particles inside the original simulation box, and using updated positions results in significantly smaller diffusion coefficients. Instead, appropriate coordinates must be used that allow the particle positions to extend beyond the confines of the simulation box.

17.2.2 Correlation functions

Assume two sets of data values x and y. Assume we wish to determine what correlation (if any) exists between them. For example, two such data sets can be the average daily temperature T_M in Minneapolis, Minnesota, and the average daily temperature T_P in Phoenix, Arizona. Other examples are the average daily temperature T_M in Minneapolis, Minnesota, and the New York Stock Exchange market closing price, or the velocity of a particle in a capillary and its distance from the wall.

A correlation coefficient C_{xy} is a numerical value that quantifies the strength of the correlation.

For M values x_i and y_i a commonly used correlation coefficient is defined as

$$C_{xy} = \frac{1}{M} \sum_{i=1}^{M} x_i y_i = \langle x_i y_i \rangle. \tag{17.2}$$

This correlation coefficient can be normalized to a value between -1 and $+1$ by dividing by the root-mean-square values of x and y:

$$C_t = \frac{\dfrac{1}{M} \sum_{i=1}^{M} x_i y_i}{\sqrt{\left(\dfrac{1}{M} \sum_{i=1}^{M} x_i^2\right)\left(\dfrac{1}{M} \sum_{i=1}^{M} y_i^2\right)}} \tag{17.3}$$

or

$$C_t = \frac{\langle x_i y_i \rangle}{\sqrt{\langle x_i^2 \rangle \langle y_i^2 \rangle}}. \tag{17.4}$$

A value of 0 indicates no correlation. An absolute value of 1 indicates strong (positive or negative correlation).

Another correlation coefficient can be defined as follows:

$$C_{xy} = \frac{\langle (x_i - \langle x \rangle)(y_i - \langle y \rangle) \rangle}{\sqrt{\langle (x_i - \langle x \rangle)^2 \rangle \langle (y_i - \langle y \rangle)^2 \rangle}}. \tag{17.5}$$

17.2.3 *Time correlation functions*

When we have data values at specific times, we can calculate the correlation between some property at some instant with another property at some later time,

$$C_{xy}(t) = \langle x(t)y(t_o) \rangle. \tag{17.6}$$

Two results that are useful are the following:

$$\lim_{t \to 0} C_{xy}(t) = \langle xy \rangle \tag{17.7}$$

and

$$\lim_{t \to \infty} C_{xy}(t) = \langle x \rangle \langle y \rangle. \tag{17.8}$$

In Chapter 12, we defined the time correlation function between any two macroscopic, mechanical quantities $A(\underline{p}(t), \underline{q}(t))$ and $B(\underline{p}(t), \underline{q}(t))$ at equilibrium as follows:

$$C_{AB}(t) = \langle A(0)B(t) \rangle - \langle A \rangle \langle B \rangle. \tag{17.9}$$

A simple program that could calculate $C_{AB}(t)$, assuming $A(t)$ and $B(t)$ are both known, for *trun* time points, could take the following form (a similar code is presented by Allen and Tildesley in their book *Computer Simulations of Liquids*, see Further reading):

```
c The correlation function will be calculated from t=0 to t=tcor,
    where tcor < trun.
c Zero out the time correlation function.
do t  =0, tcor
c(t)=0.0
enddo

do to=1, trun
tmax = min(trun, to+tcor)

do tto = to, tmax
t= tto - to
c(t) = c(t) + a(to)*b(tto)
n(t) = n(t) + 1
enddo
enddo

do t = 0, tcor
c(t) = c(t) / n(t)
enddo

end
```

With correlation functions at hand, a number of interesting properties may be computed, as discussed in Chapter 12. For example, the diffusion coefficient can be readily computed from the velocity autocorrelation function, as follows:

$$D = \frac{1}{3} \int_0^\infty \langle \underline{v}(0) \cdot \underline{v}(t) \rangle \, dt. \tag{17.10}$$

17.3 Free energy calculations

The free energy is the relevant thermodynamic property for constant temperature systems, yet absolute free energies are impossible to determine, as discussed extensively earlier in this book. Absolute free energies are thankfully also not important, since all one is interested in regarding important applications is free energy differences between states.

It is not surprising then that powerful methods have been developed for determining free energy differences from computer simulations. Here, we briefly discuss the most important classes of free energy calculations. The interested reader is also referred to the excellent book edited by Christophe Chipot and Andrew Pohorille (see Further reading).

17.3.1 Free energy perturbation methods

Let us start by considering two systems 1 and 2, with Hamiltonians H_1 and H_2, respectively. The Helmholtz free energy difference is

$$\Delta A = A_2 - A_1 = -k_B T \ln \frac{Q_2}{Q_1} \tag{17.11}$$

or

$$\Delta A = -k_B T \ln \left(\frac{\int \exp(-\beta H_2(\underline{X})) d\underline{X}}{\int \exp(-\beta H_1(\underline{X})) d\underline{X}} \right). \tag{17.12}$$

We can write

$$\Delta A = -k_B T \ln \left(\frac{\int \exp(-\beta \Delta H(\underline{X})) \exp(-\beta H_1(\underline{X})) d\underline{X}}{\int \exp(-\beta H_1(\underline{X})) d\underline{X}} \right), \tag{17.13}$$

where $\Delta H = H_2 - H_1$, is the difference between the Hamiltonians. According to the ensemble average definition, Eq. 17.13 yields

$$\Delta A = -k_B T \ln \langle \exp(-\beta \Delta H) \rangle_1. \tag{17.14}$$

If the two systems in question have the same number of particles of the same mass, the kinetic energy terms cancel out and

$$\Delta A = -k_B T \ln \langle \exp(-\beta \Delta U) \rangle_1, \qquad (17.15)$$

where $\Delta U = U_2 - U_1$ is the difference between potential energies.

The free energy perturbation method (FEP), as prescribed by Eq. 17.15, was first introduced by Robert Zwanzig in 1954. Although in principle accurate, FEP methods do not converge in practice to a correct result, unless the two end points are sufficiently close.

17.3.2 Histogram methods

The probability of observing a certain potential energy at a given temperature is given by

$$\rho(U) = \frac{\exp(-\beta U)\Omega(N, V, U)}{Z(N, V, T)}, \qquad (17.16)$$

where Ω is the microcanonical partition function and Z is the configurational integral.

A potential energy histogram can be constructed during an NVT simulation and the probability estimated from it. In practice, the number of times $N(U)$ of observing an energy within a range $U + \Delta U$ is first determined during the simulation. The probability is calculated by normalizing over all observed values of potential energy U' according to the following expression:

$$\rho(U) = \frac{N(U)}{\Delta U \sum_{U'} U'}. \qquad (17.17)$$

Given the probabilities ρ_1 and ρ_2 for two systems 1 and 2, the free energy difference is determined with the following:

$$\Delta A = -k_B T \ln \frac{\rho_2}{\rho_1}. \qquad (17.18)$$

In practice a parameter λ is defined such that λ_1 and λ_2 correspond to systems 1 and 2, respectively. The simulation methods, typically called histogram methods, proceed by calculating the ratio ρ_2/ρ_1.

17.3.3 Thermodynamic integration methods

These methods start by defining a parameter λ of the Hamiltonian, such that $H(\lambda_1)$ and $H(\lambda_2)$ are the Hamiltonians of systems 1 and 2, respectively.

The Helmholtz free energy of any system is defined as

$$A = -k_B T \ln Q(N, V, T). \tag{17.19}$$

Differentiating with respect to λ yields

$$\frac{dA}{d\lambda} = \frac{-k_B T}{Q} \frac{dQ}{d\lambda}. \tag{17.20}$$

Minor algebraic manipulation results in

$$\frac{dA}{d\lambda} = \left\langle \frac{\partial H(\lambda)}{\partial \lambda} \right\rangle_\lambda. \tag{17.21}$$

Thermodynamic integration methods proceed by integrating the average derivative of the Hamiltonian with respect to λ:

$$\Delta A = \int_{\lambda_1}^{\lambda_2} \left\langle \frac{\partial H(\lambda)}{\partial \lambda} \right\rangle_\lambda d\lambda. \tag{17.22}$$

Their significance notwithstanding, free energy calculations are difficult to implement because of sampling inaccuracies. Indeed, efforts are continuously expended by the community to improve sampling of important regions in phase space. The book *Free Energy Calculations*, see Further reading, expertly details the various free energy calculation techniques, their advantages, and challenges. The interested reader is referred to this book and the literature in it.

17.4 Further reading

1. M. P. Allen and D. J. Tildesley, *Computer Simulation of Liquids*, (London: Oxford University Press, 1989).
2. D. Frenkel and B. Smit, *Understanding Molecular Simulation: from Algorithms to Applications*, (San Diego, CA: Academic Press, 2002).
3. C. Chipot and A. Pohorille, *Free Energy Calculations*, (Berlin: Springer-Verlag, 2007).
4. R. W. J. Zwanzig, *Chem. Phys.*, **22**, 1420–1426, (1954).

17.5 Exercises

1. Prove Eq. 17.18.

2. Use an MD simulation to compute the velocity autocorrelation function of Ar at 500 K and a specific volume of 0.1736 m^3/kg. Plot the correlation function and compute the diffusion coefficient. How well does it agree with the experimental measurement (find it in the literature)? Make sure that you run with a large enough system long enough for statistics to be accurate.

18

Stochastic simulations of chemical reaction kinetics

An abundance of systems of scientific and technological importance are not at the thermodynamic limit. Stochasticity naturally emerges then as an intrinsic feature of system dynamics. In Chapter 13 we presented the essential elements of a theory for stochastic processes. The master equation was developed as the salient tool for capturing the change of the system's probability distribution in time. Its conceptual strength notwithstanding, the master equation is impossible to solve analytically for any but the simplest of systems. A need emerges then for numerical simulations either to solve the master equation or, alternatively, to sample numerically the probability distribution and its changes in time. This need resembles the need to sample equilibrium probability distributions of equilibrium ensembles.

Attempts to develop a numerical solution of the master equation invariably face insurmountable difficulties. In principle, the probability distribution can be expressed in terms of its moments, and the master equation be equivalently written as a set of ordinary differential equations involving these probability moments. In this scheme, unless severe approximations are made, higher moments are present in the equation for each moment. A closure scheme then is not feasible and one is left with an infinite number of coupled equations. Numerous, approximate closure schemes have been proposed in the literature, but none has proven adequate.

One can assume a probability distribution type, e.g., Gaussian, and eliminate high-order moments. In such a case the master equation reduces to a Fokker–Planck equation, as discussed in Chapter 13, for which a numerical solution is often available. Fokker–Planck equations, however, cannot capture the behavior of systems that escape the confines of normal distributions.

Practically, one is left with methods that sample the probability distribution and its changes in time. In this chapter, we discuss elements of numerical methods to simulate stochastic processes. We first present the work of Daniel Gillespie, who in the mid-1970s developed one of the first such methods. Gillespie's stochastic simulation algorithm (SSA) has become the foundation for many recent efforts to develop sophisticated stochastic process simulation algorithms. SSA has also become the starting point in efforts to model and simulate such diverse phenomena as gene regulatory networks and catalytic reactions.

18.1 Stochastic simulation algorithm

Let us consider a well-mixed volume V, containing N distinct chemical species S_i participating in M chemical reactions. The system is fully described by a state vector $\underline{X}(t) = (X_1(t), \ldots, X_N(t))$, which contains the number of molecules of each species at any time t.

A stoichiometric $M \times N$ matrix, $\underline{v}$, can be defined with elements v_{ij} representing the change in the number of molecules of the ith species caused by the jth reaction.

Let us define the propensity for each reaction with $a_j(\underline{X}(t))dt$ as the probability that the jth reaction occurs in a small time interval $[t, t + dt]$.

Propensities may be calculated using different rate laws such as mass action or Michaelis–Menten kinetics. Using mass action kinetics, the probabilistic reaction rates are calculated given the macroscopic (deterministic) rate constant k_j and the corresponding reaction form and law, as discussed in Chapter 13.

In general, the propensity of the jth reaction can be calculated using the equation

$$a_j\left(\underline{X}(t)\right) = c_j h_j\left(\underline{X}(t)\right), \tag{18.1}$$

where h_j is the number of distinct combinations of the reacting species and c_j is the average specific reaction propensity for the jth reaction. As discussed in Chapter 13, c_j is related to the reaction rate constant k_j.

The original stochastic simulation algorithm developed by Gillespie is also known as the Direct Method SSA variant. At the heart of the Direct Method is determination of the time, τ, that the next reaction will occur in the system and identification of the reaction, μ, that will occur next. In practice, τ and μ are determined as follows:

$$\tau = \frac{1}{a} \ln\left(\frac{1}{r_1}\right) \tag{18.2}$$

and

$$\sum_{j=1}^{\mu-1} a_j < r_2 a \le \sum_{j=1}^{\mu} a_j, \tag{18.3}$$

where a is the sum of all reaction propensities and r_1 and r_2 are uniform random numbers.

Briefly, the algorithm first determines when the next reaction will occur based on the sum of probabilistic reaction rates, using Eq. 18.2. The method then establishes which of the M reaction channels will indeed occur, given the relative propensity values. The μth reaction occurs when the cumulative sum of the μ first terms becomes greater than $r_2 a$ (Eq. 18.3).

Implementation of Eqs. 18.2 and 18.3 samples the probability distribution of these events, $P(\tau, \mu)$, which is expressed as follows:

$$P(\tau, \mu) = a_\mu \exp\left(-\tau \sum_j a_j\right). \tag{18.4}$$

Subsequently the reaction propensities are updated, since the system has "jumped" to a neighboring state and the process is repeated until the system reaches the desired end time point.

Gillespie's algorithm numerically reproduces the solution of the chemical master equation, simulating the individual occurrences of reactions. This type of description is called a jump Markov process, a type of stochastic process. A jump Markov process describes a system that has a probability of discontinuously transitioning from one state to another. This type of algorithm is also known as kinetic Monte Carlo. An ensemble of simulation trajectories in state space is required to accurately capture the probabilistic nature of the transient behavior of the system.

18.2 Multiscale algorithms for chemical kinetics

The stochastic simulation algorithm simulates every single reacting event, which renders the algorithm accurate, but at the same time computationally demanding. It becomes especially unfavorable when there is a large number of reactions occurring with high frequency. There have been numerous attempts to improve the efficiency of SSA within the last ten years.

Gibson and Bruck improved the performance of SSA by resourcefully managing the need for random numbers, creating the Next Reaction variant of SSA. Cao and co-workers optimized the Direct Reaction variant of the SSA, proving that for certain systems this approach is more efficient than the Next Reaction variant.

A number of other mathematically equivalent approximations to the SSA have been proposed, aiming to balance efficiency with complexity, reviewed in 2007 by Gillespie (see Further reading).

As a first approximation, when developing a modeling formalism for a reacting system, the state space associated with chemical reaction networks can be partitioned into regions that depend on the nature of the system. Doing so is valuable since it allows one to evaluate which approximations are reasonable, and provides a comprehensive picture for the segue between the regions where models must be solved with stochastic methods and where ordinary differential equations can be used.

The essential elements to consider for spatially homogeneous systems are whether the species populations are continuous or discrete and whether the chemical reaction events are rare or frequent. In Fig. 18.1 a schematic is presented of the modeling regions, with the x-axis measuring the number of molecules of reacting species and the y-axis

Figure 18.1 Regimes of the problem space for multiscale stochastic simulations of chemical reaction kinetics. The x-axis represents the number of molecules of reacting species, x, and the y-axis measures the frequency of reaction events, Λ. The threshold variables demarcate the partitions of modeling formalisms. In area I, the number of molecules is so small and the reaction events are so infrequent that a discrete-stochastic simulation algorithm, like the SSA, is needed. In contrast, in area V, which extends to infinity, the thermodynamic limit assumption becomes valid and a continuous-deterministic modeling formalism becomes valid. Other areas admit different modeling formalisms, such as ones based on chemical Langevin equations, or probabilistic steady-state assumptions.

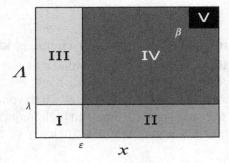

measuring the frequency of reaction events. The threshold variables demarcate the different partitions. Their definitions are as follows:

1. β is the threshold value of the ratio between stochastic variation and deterministic behavior.
2. λ is the threshold for the number of reaction events per time beyond which a reaction is considered "fast."
3. ε is the threshold for the molecules number beyond which the population can be treated continuously.

It should be noted that the state-space will be two-dimensional only if there is one reactant and one reaction. With N reactants and M reactions, the problem-space will be $(M \times N)$-dimensional.

18.2.1 Slow-discrete region (I)

In the slow-discrete region, species populations are so small that a difference in one molecule is significant. Thus, the species must be represented with integer variables, and only discrete changes in the populations are allowed. Additionally, the reaction events are very rare. Typically, these events correspond to significant changes in the rest of the system, and are the ones of interest in multiscale simulations. Because the rates are low, the rest of the system is usually able to relax to some steady-state manifold. This region should be treated with the highest accuracy method available, the numerical solution of the master equation, or alternatively the exact stochastic simulation algorithm.

18.2.2 Slow-continuous region (II)

In the slow-continuous region, species populations are sufficiently large such that molecular populations may be assumed to be continuous. Furthermore, since the populations are quite large with respect to the number of reaction occurrences, each occurrence of a region II reaction does not significantly change its species concentrations. Therefore, we are able to skip the simulation of many reaction occurrences without significantly affecting the accuracy of the large populations. If this region is to be approximated, the number of reaction events in a time frame can be taken from a Poisson distribution.

18.2.3 Fast-discrete region (III)

In the fast-discrete region, molecular populations are small enough that they must be calculated with discrete variables. However, the reactions happen frequently enough that exact simulations of this regime are slow.

This is the region that is typically referred to as "stiff." Because of the high reaction rates, the species populations should arrive at steady-state distributions quickly, with respect to the occurrences of slower reactions. Because the number of molecules is small, we cannot assume any specific shape for the species distributions. If we can quickly compute the molecular distributions of species affected, then we can skip many reaction steps by sampling from this equilibrium distribution.

18.2.4 Fast-continuous stochastic region (IV)

In the fast-continuous region, species populations can be assumed to be continuous variables. Because the reactions are sufficiently fast in comparison to the rest of the system, it can be assumed that they have relaxed to a steady-state distribution. Furthermore, because of the frequency of reaction rates, and the population size, the population distributions can be assumed to have a Gaussian shape. The subset of fast reactions can then be approximated as a continuous time Markov process with chemical Langevin Equations (CLE). The CLE is an Itô stochastic differential equation with multiplicative noise, as discussed in Chapter 13.

18.2.5 Fast-continuous deterministic region (V)

As the number of reaction firings in a time step increases, the ratio of the noise term to the non-stochastic reaction rate in the CLE decreases. In the limit of fast reactions, noise vanishes, and reactions can be modeled with ordinary differential equations. Note that this modeling formalism area extends to the thermodynamic limit.

18.3 Hybrid algorithms

In general, approximations that may improve the computational efficiency of the SSA can be classified into two major categories, time-leaping methods and system-partitioning methods. The time-leaping methods depend on the assumption that many reacting events can occur in a time period without significantly changing the reaction probabilities, i.e., the change in the state of the system minimally impacts the reaction propensities. This group includes the explicit and implicit tau-leaping algorithms, which use Poisson random variables to compute the reacting frequency of each reacting event in a given time interval. The main drawback of the tau-leaping approximation is that it becomes inaccurate when a significant number of critical reactions (reactions where even a single reacting event significantly impacts the reaction propensities) are included in a single leap such that the reaction propensities change excessively or some molecular populations become negative. Concerns have been addressed by adaptively restricting the size of each individual leap. While these recent versions appear to be more rigorous they still are insufficient when small numbers of reacting molecules result in dramatic changes in propensity functions.

The second approach to speeding up the SSA involves separating the system into slow and fast subsets of reactions. In these methods, analytical or numerical approximations to the dynamics of the fast subset are computed while the slow subset is stochastically simulated. In one of the first such methods, Rao and Arkin (see Further reading) applied a quasi-steady-state assumption to the fast reactions and treated the remaining slow reactions as stochastic events.

Hybrid methods that combine modeling formalisms are receiving considerable interest (see Further reading). Puchalka and Kierzek partitioned the system into slow and fast reaction subsets, with the first propagated through the Next Reaction variant and the latter through a Poisson (tau-leaping) distribution. Haseltine and Rawlings also partition the system into slow and fast reactions, representing them as jump and continuous Markov processes respectively. Both aforementioned hybrid methods suffer when it comes to implementation issues, making them slower or inaccurate.

Salis and Kaznessis separated the system into slow and fast reactions and managed to overcome the inadequacies and achieve a substantial speed up compared to the SSA while retaining accuracy. Fast reactions are approximated as a continuous Markov process, through Chemical Langevin Equations (CLE), discussed in Chapter 13, and the slow subset is approximated through jump equations derived by extending the Next Reaction variant approach.

Except for those two major categories there are also other approaches that cannot be classified as one or the other group. Such methods include the equation free probabilistic steady state approach of Salis and Kaznessis. In this methodology reactions are partitioned into slow/discrete and fast/discrete subsets and the future states are predicted through the sampling of a quasi-steady-state marginal distribution. A similar approach is followed in the work of Samant and Vlachos. Erban and co-workers also proposed an equation-free numerical technique in order to speed up SSA. SSA is used to initialize and estimate probability densities and then standard numerical techniques propagate the system. In a different approach, Munsky and Kammash used projection formalism to truncate the state space of the corresponding Markov process and then directly solve or approximate the solution of the CME.

18.4 Hybrid stochastic algorithm

Salis and Kaznessis proposed a hybrid stochastic algorithm that is based on a dynamical partitioning of the set of reactions into fast and slow subsets. The fast subset is treated as a continuous Markov process governed by a multidimensional Fokker–Planck equation, while the slow subset is considered to be a jump or discrete Markov process governed by a CME. The approximation of fast/continuous reactions as a continuous Markov process significantly reduces the computational intensity and introduces a marginal error when compared to the exact jump Markov simulation. This idea becomes very useful in systems where reactions with multiple reaction scales are constantly present.

18.4.1 System partitioning

Given a stochastic chemical kinetics model, the set of reactions is dynamically apportioned into two subsets, the fast/continuous and slow/discrete reactions. Now M is the sum of the fast M^{fast} and slow M^{slow} reactions, respectively. Propensities are also designated as fast $\underline{a}^{\text{f}}$ and slow $\underline{a}^{\text{s}}$.

For any reaction to be classified as fast, the following two conditions need to be met:

- The reaction occurs many times in a small time interval.
- The effect of each reaction on the numbers of reactants and products species is small, when compared to the total numbers of reactant and product species.

In equation form, these criteria can be expressed as follows:

$$\text{(i)} \quad a_j\big(\underline{X}(t)\big) \geq \lambda \gg 1,$$
$$\text{(ii)} \quad X_i(t) > \varepsilon |\nu_{ji}|, \tag{18.5}$$

where the ith species is either a product or a reactant in the jth reaction.

The two parameters λ and ε define respectively the numbers of reactions occurring within time Δt and the upper limit for the reaction to have a negligible effect on the number of molecules of the reactants and products. This approximation becomes valid when both λ and ε become infinite, i.e., at the thermodynamic limit. In practice, typical values for

λ and ε may be around 10 and 100, respectively. Obviously the conditions must be evaluated multiple times within a simulation since both the propensities and the state of the system may change over time. This practically means that it is possible for one reaction to change subsets, i.e., fast or slow, within an execution.

18.4.2 *Propagation of the fast subsystem – chemical Langevin equations*

The fast subset dynamics are assumed to follow a continuous Markov process description and therefore a multidimensional Fokker–Planck equation describes their time evolution. The multidimensional Fokker–Plank equation more accurately describes the evolution of the probability distribution of only the fast reactions. The solution is a distribution depicting the state occupancies. If the interest is in obtaining one of the possible trajectories of the solution, the proper course of action is to solve a system of chemical Langevin equations (CLEs).

A CLE is an Itô stochastic differential equation (SDE) with multiplicative noise terms and represents one possible solution of the Fokker–Planck equation. From a multidimensional Fokker–Planck equation we end up with a system of CLEs:

$$dX_i = \sum_{j=1}^{M^{\text{fast}}} v_{ji} a_j(\underline{X}(t)) dt + \sum_{j=1}^{M^{\text{fast}}} v_{ji} \sqrt{a_j(\underline{X}(t))} dW_j, \quad (18.6)$$

where a_j, v_{ji} are the propensities and the stoichiometric coefficients respectively and W_j is a Wiener process responsible for the Gaussian white noise.

It is beyond the scope of this book to present methods for the numerical integration of stochastic differential equations. The interested reader is referred to the excellent book by Kloeden and Platen (see Further reading).

18.4.3 *Propagation of the slow subsystem – jump equations*

On the other hand, the time evolution of the subset of slow reactions is propagated in time using a slight modification of the Next Reaction variant of SSA, developed by Gibson and Bruck. A system of differential jump equations is used to calculate the next jump of any slow reaction. The jump equations are defined as follows,

$$dR_j(t) = a_j^s(\underline{X}(t)) dt,$$
$$R_j(t_0) = \log(r_j), \quad j = 1, \ldots, M^{\text{slow}} \quad (18.7)$$

where R_j denotes the residual of the jth slow reaction, a_j^s are the propensities of only the slow reactions and r_j is a uniform random number in the interval $(0, 1)$. Equation 18.7 depicts the rate at which the reaction residuals change. Note that the initial conditions of all R_j are negative. The next slow reaction occurs when the corresponding residual value makes a zero crossing, from negative to positive values.

Equations 18.7 are also Itô differential equations even though they do not contain any Wiener process, because the propensities of the slow reactions depend on the state of the system, which in turn depends on the system of CLEs. Due to the coupling between the system of CLEs and the differential jump equations, a simultaneous numerical integration is necessary. If there is no coupling between fast and slow subsets or there are only slow reactions the system of differential jump equations simplifies to the Next Reaction variant.

The method can be further sped up by allowing more than one zero crossing, i.e., more than one slow reaction, to occur in the time it takes the system of CLEs to advance by Δt. Though this is an additional approximation contributing to the error introduced by the approximation of the fast reactions as continuous Markov processes, it results in a significant decrease in simulation times. The accuracy depends on the number of slow reactions allowed within Δt and decreases as the number increases.

18.5 Hy3S – Hybrid stochastic simulations for supercomputers

Hy3S, or Hybrid Stochastic Simulation for Supercomputers, is an open sourced software package written for the development, dissemination, and productive use of hybrid stochastic simulation methods. The goal of the software is to allow users to utilize the hybrid stochastic simulation algorithms described in the previous section and to simulate large, realistic stochastic systems.

The software package is freely available at hysss.sourceforge.net and at statthermo.sourceforge.net and includes multiple different hybrid stochastic simulation methods implemented in FORTRAN 95 and a simple MATLAB (Mathworks) driven graphics user interface. It uses the NetCDF (Unidata) interface to store both model and solution data in an optimized, platform-independent, array-based, binary format.

A detailed description of how to download the software and related libraries can be found on the website hysss.sourceforge.net. A thorough discussion of the methods and the GUI, along with the presentation of several implementation examples, can be found in the 2006 manuscript by Salis and co-workers (see Further reading).

18.6 Multikin – Multiscale kinetics

Multikin is a suite of software for modeling and simulation of arbitrary chemical reaction networks. At the core of Multikin lies Hy3S for simulating the dynamic behavior of reacting systems using hybrid stochastic-discrete and stochastic-continuous algorithms. Multikin evolved from SynBioSS, which is a software suite (www.synbioss.org), originally developed to model biomolecular systems. Multikin is available under the GNU General Public License at the textbook's website, http://statthermo.sourceforge.net.

Multikin is implemented in Python (http://www.python.org) using GTK+ (http://www.gtk.org) to provide a user-friendly graphical interface, without the need for proprietary software like MATLAB. These choices enable cross-platform deployment of the same code. In Multikin, users can create a reaction network model by adding reactions using a drop-down menu of various reaction rate laws (there are tutorials in the textbook's website). Multikin uses libSBML also to read models specified in SBML and can also read models created directly for Hy3S as NetCDF files. Next, the user may modify the loaded model. The user can subsequently specify simulation parameters. Typically, this is merely the amount of simulation time and the number of stochastic trajectories to be sampled (stochastic simulation requires multiple samples to construct population distributions). Many other options are available, such as approximation cutoffs, integration schemes, etc., but setting these is typically unnecessary, as defaults for stable simulations are provided. After setting parameters, the user then may either run the simulation locally or export a NetCDF file appropriate for the supercomputer simulator. If conducted locally, the results may be exported as either an ASCII comma-separated-value (CSV) file for import into any spreadsheet program, or as a NetCDF file appropriate for MatLab.

18.7 Further reading

1. P. E. Kloeden and E. Platen, *Numerical Solution of Stochastic Differential Equations*, (Berlin: Springer, 1995).
2. C. W. Gardiner, *Handbook of Stochastic Methods: for Physics, Chemistry and the Natural Sciences*, (Berlin: Springer, 2004).

3. N. G. Van Kampen, *Stochastic Processes in Physics and Chemistry*, (Amsterdam: North-Holland Personal Library, 1992).

4. D. T. Gillespie, *Markov Processes: An Introduction for Physical Scientists*, (San Diego, CA: Academic Press, 1991).

5. H. Salis and Y. Kaznessis, Numerical simulation of stochastic gene circuits, *Computers & Chem. Eng.*, **29** (3), 577–588, (2005).

6. H. Salis and Y. Kaznessis, An equation-free probabilistic steady state approximation: Dynamic application to the stochastic simulation of biochemical reaction networks, *J. Chem. Phys.*, **123** (21), 214106, (2005).

7. H. Salis, V. Sotiropoulos, and Y. Kaznessis, Multiscale Hy3S: Hybrid stochastic simulations for supercomputers, *BMC Bioinformatics*, **7** (93), (2006).

8. D. T. Gillespie, A general method for numerically simulating the stochastic time evolution of coupled chemical reactions, *J. Comput. Phys.*, **22**, 403–434, (1976).

9. M. A. Gibson and J. Bruck, Efficient exact stochastic simulation of chemical systems with many species and many channels, *J. Phys. Chem. A*, **104**, 1876–1889, (2000).

10. Y. Cao, H. Li, and L. Petzold, Efficient formulation of the stochastic simulation algorithm for chemically reacting systems, *J. Chem. Phys.*, **121**, 4059–4067, (2004).

11. E. L. Haseltine and J. B. Rawlings, Approximate simulation of coupled fast and slow reactions for stochastic chemical kinetics, *J. Chem. Phys.*, **117**, 6959–6969, (2002).

12. J. Puchalka and A. M. Kierzek, Bridging the gap between stochastic and deterministic regimes in the kinetic simulations of biochemical reaction networks, *Biophys. J.*, **86**, 1357–1372, (2004).

13. D. T. Gillespie, The chemical Langevin equation, *J. Chem. Phys.*, **113**, 297–306, (2000).

14. C. V. Rao and A. P. Arkin, Stochastic chemical kinetics and the quasi-steady-state assumption: application to the Gillespie algorithm, *J. Chem. Phys.*, **118**, 4999–5010, (2003).

15. Samant Asawari, A. Ogunnaike Babatunde and G. Vlachos Dionisios, A hybrid multiscale Monte Carlo algorithm (HyMSMC) to cope with disparity in time scales and species populations in intracellular networks, *BMC Bioinformatics*, **8**, 175, (2007).

16. B. Munsky and M. Khammash, The finite state projection algorithm for the solution of the chemical master equation, *J. Chem. Phys.*, **124**, 044104, (2006).

17. D. T. Gillespie, Stochastic simulation of chemical kinetics, *Ann. Rev. Phys. Chem.*, **58**, 35–55, (2007).

18. R. Erban, I. Kevrekidis, D. Adalsteinsson, and T. Elston, Gene regulatory networks: A coarse-grained, equation-free approach to multiscale computation, *J. Chem. Phys.*, **124**, 084–106, (2006).

18.8 Exercises

1. For the four chemical reaction networks below, write the chemical master equation. Give the solution for the first two systems:

a)
$$A_1 \underset{k_{-1}}{\overset{k_1}{\rightleftharpoons}} B_2.$$

b)
$$2A_1 \underset{k_{-1}}{\overset{k_1}{\rightleftharpoons}} A_2 + A_3.$$

c)
$$A_1 + A_2 \overset{k_1}{\rightarrow} A_3 + A_4,$$
$$A_3 + A_2 \overset{k_2}{\rightarrow} 2A_4,$$
$$A_1 + A_3 \overset{k_3}{\rightarrow} 2A_3 + A_5,$$
$$2A_3 \overset{k_4}{\rightarrow} A_1 + A_6,$$
$$A_5 \overset{k_5}{\rightarrow} 2A_2.$$

d)
$$2A_1 \overset{k_1}{\rightarrow} A_3,$$
$$A_1 + A_2 \overset{k_2}{\rightarrow} A_4,$$
$$A_4 + A_5 \overset{k_3}{\rightarrow} A_2.$$

2. Simulate these four reaction networks with Hy3S or Multikin. Vary the kinetic parameter values as well as the initial conditions. Verify that for linear systems the fluctuations scale with the inverse of the square root of the numbers of molecules. Comment on the need for stochastic simulations away from the thermodynamic limit.

◼︎◼︎◼︎ Appendix A ◼︎◼︎◼︎

Physical constants and conversion factors

A.1 Physical constants

Table A.1 *Physical constants.*

Quantity	Symbol	Value
Atomic mass unit	amu	1.6605×10^{-27} kg
Avogadro's number	N_A	6.0225×10^{23} molecules mole^{-1}
Boltzmann's constant	k_B	1.3806×10^{-23} m^{2}kgs^{-2}K^{-1}
Elementary charge	e	1.602×10^{-19} C
Gas constant	R	8.3143 J K^{-1} mol^{-1}
Planck's constant	h	6.626×10^{-34} m^{2}kgs^{-1}

A.2 Conversion factors

Table A.2 *Conversion factors.*

To convert from	To	Multiply by
Atmosphere	newton meter^{-2}	1.013×10^5
Bar	newton meter^{-2}	1.000×10^5
British thermal unit	joule	1.055×10^3
Calorie	joule	4.184
Dyne	newton	1.000×10^{-5}
Electron volt	joule	1.602×10^{-19}
Erg	joule	1.000×10^{-7}
Inch	meter	2.540×10^{-2}
Lbf (pound-force)	newton	4.448
Torr	newton meter^{-2}	1.333×10^2

Appendix B

Elements of classical thermodynamics

B.1 Systems, properties, and states in thermodynamics

A system is the part of the universe we are interested in. The rest of the universe is called surroundings. The system volume, V, is well defined and the system boundary is clearly identified with a surface.

Systems are described by their mass M and energy E. Instead of using mass M, a system may be defined by the number of moles, N/N_A, where N is the number of molecules and N_A is Avogadro's number.

There are three kinds of system in thermodynamics:

1. *Isolated systems*. In these systems, there is no mass or energy exchange with the surroundings.
2. *Closed systems*. In these systems, there is no mass exchange with the surroundings. Energy can flow between the system and the surroundings as heat, Q, or work, W. Heat is the transfer of energy as a result of a temperature difference. Work is the transfer of energy by any other mechanism.
3. *Open systems*. In these systems, mass and energy may be exchanged with the surroundings.

A thermodynamic state is a macroscopic condition of a system prescribed by specific values of thermodynamic properties.

The following thermodynamic properties define the state of a system:

- Temperature, T,
- Pressure, P,
- Internal energy, E,
- Enthalpy, H,
- Entropy, S,
- Gibbs free energy, G,
- Helmholtz free energy, A,
- Chemical potential, μ.

The following relations between thermodynamic properties are always true for closed or isolated systems:

$$H = E + PV, \tag{B.1}$$

$$A = E - TS, \tag{B.2}$$

$$G = H - TS. \tag{B.3}$$

The number of thermodynamic properties that can uniquely define the state of a system is determined by the Gibbs phase rule:

$$F = C - P + 2, \tag{B.4}$$

where F is the number of necessary thermodynamic properties, C is the number of molecular components in the systems, and P is the number of phases in the system (e.g., gas, liquid, or solid). For single-component, single-phase systems, only two thermodynamic variables are necessary to define the state of the system. This means that the values of the rest of the thermodynamic properties are determined when the values of any two thermodynamic properties are defined.

An isolated or closed system is at an equilibrium state if there are no internal gradients in pressure, temperature, and chemical potential.

B.2 Fundamental thermodynamic relations

The first law of thermodynamics for closed systems can be written in differential form as follows:

$$dE = \delta Q + \delta W, \tag{B.5}$$

where δQ and δW are small amounts of heat and work exchanged between the system and the environment.

If the heat is exchanged slowly enough for the process to be considered reversible, then the second law of thermodynamics postulates that

$$dS = \delta Q / T, \tag{B.6}$$

where dS is an infinitesimal change in the entropy of the system.

If the work is only associated with a change in the volume of the system, dV, we may write:

$$\delta W = -PdV. \tag{B.7}$$

The change in the energy is then

$$dE = TdS - PdV. \tag{B.8}$$

This is the first of four fundamental relations between thermodynamic properties. Combining Eqs. B.1, B.2, B.3, and B.8 results in the remaining three fundamental thermodynamic relations:

For the enthalpy

$$dH = SdT + VdP. \tag{B.9}$$

For the Helmholtz free energy

$$dA = -TdS - VdP. \tag{B.10}$$

For the Gibbs free energy

$$dG = SdT - VdP. \tag{B.11}$$

The fundamental relations can be modified for open systems. If there are dN_j molecules exchanged between the system and its environment for each of M molecular components, with $j = 1, \ldots, M$, then

$$dE = TdS - PdV + \Sigma_{j=1}^{M} \mu_j dN_j, \tag{B.12}$$

$$dH = SdT + VdP + \Sigma_{j=1}^{M} \mu_j dN_j, \tag{B.13}$$

$$dA = -TdS - VdP + \Sigma_{j=1}^{M} \mu_j dN_j, \tag{B.14}$$

$$dG = SdT - VdP + \Sigma_{j=1}^{M} \mu_j dN_j. \tag{B.15}$$

Equations B.8, B.10, and B.11 reveal important closed system equilibrium criteria:

1. For constant E and V ($dE = 0$ and $dV = 0$), the entropy of the system attains a maximum value ($dS = 0$; also $d^2 S < 0$, whose proof is beyond the scope of this appendix).
2. For constant T and V ($dT = 0$ and $dV = 0$), the Helmholtz free energy of the system attains a minimum value.
3. For constant T and P ($dT = 0$ and $dP = 0$), the Gibbs free energy of the system attains a minimum value.

Index

Printed in the United States
By Bookmasters